Statistics
A User Friendly Guide
(Especially for the Mathematically Challenged)

Gerald C. Swanson, Ph.D.

GREYHERON Press
Edmonds, Washington

Library of Congress Control Number: 2002094650

ISBN 0-9724589-0-5

Printed in the United States of America
October 2002

GREYHERON Press
PO Box 1644
Lynnwood, Washington 98046
information@greyheronpress.com
www.greyheronpress.com

I dedicate this book to my wife Terri in thanks for her continued support of my desire to teach. I also dedicate it to the many students who have contributed to its contents and who have continually challenged me to clearly communicate its contents.

Contents

Introduction – How to Use This Book
READ THIS FIRST!

This book results from teaching statistics for over seven years to students (Behavioral Scientists and Psychological Counselors) who are traditionally not well-grounded in mathematics. The content, sequence and techniques have evolved from my commitment that ***no student is left behind*** in understanding this subject matter. Much of the content is an effort to put on the written page the many experiential techniques that have worked so well in the classroom.

Basic Statistics (Pages 1 – 98)

- Study this material in sequence
- Use a learning partner
- Do all the exercises – then read the explanations

Learning statistics is a cumulative process in which new knowledge builds on prior knowledge. The sequence of topics in this section has been proven through long experience to build this cumulative knowledge well. For your most effective learning, start at the beginning and proceed sequentially through the text. The exercises are a critical part of the learning process. They are designed to provide the key experiences to both communicate and solidify the learning of key subject matter. Often new material is introduced in the explanations following the exercises. *For maximum learning, all exercises must be performed.* If at all possible, find a learning partner who will interact with you as you go through this material at a shared pace. Much of the learning will occur through sharing the different perspectives and understandings you and your learning partner develop.

This basic introduction to statistics does not cover the actual performance of statistical tests. The material emphasizes conceptual understanding while requiring a minimum amount of actual calculations. I recommend that you acquire a calculator to use while learning basic statistics. A "five-function" calculator (adds, subtracts, multiplies, divides and takes a square root) will meet your simple calculation needs.

Advanced Statistics (Pages 99 – 158)

This material is appropriate for an advanced level statistics course. It primarily addresses the performance of statistical tests of significance. I have employed it for instructing Masters level students in the Behavioral Sciences who also may be poorly grounded in mathematics. In such instance, the Basic Statistics materials are for review and reference to assure that necessary fundamentals are present in performing these tests. Again the exercises are a primary method for developing and reinforcing the learning from the text. In the performance of practice calculations, nothing serves for learning as well as actually performing the calculations.

Statistics A User Friendly Guide
(Especially for the Mathematically Challenged)

1 The Fundamental Statistical Process

Before Beginning:

This is a guide to aid you in understanding basic and advanced statistics. Before beginning, it is helpful to examine your experience with statistics or other types of mathematics.

- How did you feel about your actual learning in these previous studies?
- How did you feel about how others (teachers and other students) treated you?
- How would you like to progress and experience your learning in this course of study?

Many statistics students, who have previously been very reluctant to take any mathematics, have been able to learn this subject matter very successfully. Develop an expectation that you can do so also. Two actions you can take will particularly contribute to this success:

1. Set a purposeful pace for yourself in using this book, including completion of all the exercises. (Just as in physical conditioning, exercise builds capacity.)
2. If possible, work with a learning partner. Different perspectives and methods of creating understanding aid the learning process. Expand your positive self-expectations to your learning partner. Create a joint commitment to your mutual success.

What is Statistics?

Statistics is a number of things to a number of people: a pain, a joy, a tool, a puzzle. But foremost, statistics is a *simplification of reality*. This definition is often counter-intuitive because most students view statistics as being very complicated. It is important to understand from the outset that the whole purpose of statistics is to simplify a very complex reality to a representation by a few numbers. This guide will continually point out that the fundamental statistical process is simplification. Some students may already inherently grasp that statistics can (over) simplify reality and resent this simplification that statistics produces. Such resentment may create an inner resistance to learning statistics and, unless consciously addressed may block learning. We have all heard the phrase "He or she is being reduced to a statistic" and understood it as a negative statement. By consciously addressing any such concerns on your part, you may be able to move beyond them and appreciate the value and beauty of the statistical process.

Exercise 1.1
Take a moment and reflect on your understanding of statistics at this point. Write a few comments on the space below. You will be asked to repeat this exercise at several points as a method of tracking your progress in learning statistics.

____

____

____

____

____

Another way to think about statistics is as a foreign language. There are many different ways to study foreign languages, and there are many different ways to study statistics. This user-friendly approach to learning statistics is to learn the underlying logic and thought processes. You will do minimal calculations as a way to help you make sure that you have understood the key concepts. One complicating aspect of statistics as a language is that it uses a lot of the same words as conversational English. This can create confusion because when these common everyday English words are used statistically, they have very exact meanings that must not be confused with the more varied everyday meaning. Whenever a new statistical term is introduced in this guide, it will be highlighted in <u>*underlined italics*</u> to draw your attention.

Another way of understanding statistics is that it is a branch of <u>*applied mathematics*</u>. Let's examine this aspect of statistics starting with the word "applied". Statistics is not a theoretical mathematical subject. It is a practical collection of very powerful tools for measurement and representation of reality. These tools are applied for a purpose, to accomplish useful actions. As such it is always legitimate to ask the question "So What?" Answering this question details how the application of statistics is of value to you.

As a branch of mathematics, statistics is a set of mathematical rules and regulations for generating the correct answers. This emphasis on the correct answers occasionally results in producing formidable anxiety that acts as a block to learning. The mathematics underlying statistical applications is cumulative, that is new learning builds on prior learning. This guide builds from the simplest concepts and mathematics towards the more complex. It is important to study this material in the sequence presented, particularly if you are prone to any anxieties around this subject matter. Long experience in the delivery of this material has created this sequence of presentation that builds cumulatively on conceptual understanding and mathematical competency.

The Fundamental Statistical Processes: Describing Reality

The first applications of statistics occurred very early in human history. The most powerful driver for describing and simplifying reality was the need for rulers to raise taxes from those ruled. In order to facilitate this action, the taking of census has occurred since pre-history. (Some of the earliest written records found are summaries of early census results.) The value of the census has not declined with time. In the year 2000, the United States performed a census of its population at a cost in excess of $5 Billion. Such census activities are intended to produce a total description (count) of the population.

The process of using statistics to describe the totality of a chunk of reality is called *Descriptive Statistics*. Of course this description of the totality focuses on specific aspects of interest and ignores other aspects of the reality. Thus the examples of census cited above focus on actual counts of human populations often broken into geographic areas of interest. Let's create an example description as a way of furthering our understanding of descriptive statistics.

Exercise 1.2

Spend some time (in collaboration with your learning partner) to describe your immediate surroundings. Create a description of 20 to 30 words:

____

____

____

____

____

The exercise you just completed is akin to the fundamental statistical process of taking a chunk of reality and describing it with only one or two numbers. When you describe any aspect of reality that has more than one component, what is the descriptive process? There are really only two ways that you can meaningfully describe reality - describe the similarities, and - describe the differences. The similarities are the common bonds, the things that make us or them alike. The differences are the things that separate and differentiate, the things that allow us to know the differences. Statistics as a simplification of reality also describes using just two numbers, one to show similarities and one to show differences. Go back over the description you created in Exercise 1.2 and underline the words that relate to how the reality you described is similar. Circle the words that relate to how the reality you described is different. If you have difficulty discriminating between these in your description, ask for help from your learning partner. It is difficult to imagine an effective description that only relates how reality is similar, or only relates how it is different. An effective description addresses both similarities and differences.

In order to describe a chunk of reality you must be able to separate that chunk from the rest of reality. The boundary must be distinct and clear. When we have made such a distinction we must be able to describe the distinction to others so they can separate the *same* chunk of reality. Statistics gives that separated chunk a name; it is called a <u>population</u>. The use of this word to designate a group of people that we will describe is pretty clear since it fits with our every day language usage. The statistical usage of population, however, extends far beyond the normal everyday usage. In statistics, *a population is any chunk of reality which you can clearly distinguish from the rest of reality and which you describe statistically.* It can be a population of the dirt removed from beneath pier 21 on the Seattle waterfront. It can be a population of all emperor penguins living on the Antarctic ice pack. It can be a population of all water flowing through a specific river basin during a specific period of time. The definition of a population needs to be an *operational definition*. An operational definition tells how to accomplish the thing being defined. (In statistics most definitions are Operational Definitions - definitions that tell you how to accomplish the thing being defined as opposed to theoretical definitions.) In the case of developing an operational definition of a population, the definition tells how to draw the lines and establish the boundaries clearly distinguishing your chunk of reality, your population, from the rest of reality. The necessary condition is that the population description has such clarity that anyone else could repeat the process and re-create (at least theoretically) the population.

Exercise 1.3
Create an operational definition of a population relevant to your work or study area (again in collaboration with your learning partner). Make sure your definition is sufficiently detailed that another person could (at least theoretically) recreate the population just from your definition:

____

____

____

____

____

____

If you are working with a learning partner, exchange the population definitions created in exercise 1.3. Examine your partner's definition to see if it is sufficiently clear and explicit that you could (at least theoretically) recreate the population. Does the description make clear the boundaries between the population described and the rest of reality?

Is it clear in location (where)? Is it clear in time (when)? Is it clear in extent (how much)? Any given population may not be static, i.e. the water flowing in a river basin through a specific period of time. In such a case the population could not in actuality be recreated, but the critical issue is that it the definition is sufficiently specific that theoretically it could be.

Once you have defined your population, the next step is to prepare to describe it. The whole purpose of statistics is to simplify reality into something that can be readily understood and worked with. As you will be learning throughout the rest of this guide there are really only two aspects to that description of reality: the similarities and the differences. Before we can develop statements of similarities and differences, there is one more step that we must take: develop a way to measure the reality and produce numbers that represent it.

2 Measurement

Types of Measures

In order to make any kind of statistical statement about a population it is necessary to deal with numbers. The simplest kind of numbering or measuring process is to count. A little more sophisticated process is to create a number (called a measure) for each member of the population that expresses some key characteristic of that population member. This chapter deals with the numbers we use to count or measure, and how we generate and use them. There are four kinds of measures used to generate numbers (or data) for use in statistics. These four measures have different levels of sophistication and therefore are in a hierarchy of the information that they convey and how they can be used statistically. The measures are: *nominal*, *ordinal*, *interval* and *ratio*. Returning to the language of statistics, these four types of measures are called: *levels of scaling*. (*Scaling is the process of converting properties of population elements into numbers reflecting either counts or other numerical evaluations*.) So What? The distinctions between these levels of scaling are very important. Just as the ingredients you use dictate what kind of cooking you can do. (You can't make a salad from dough, yeast, salt and water!) So the properties of these levels of scaling affect how the numbers or data generated are used in all aspects of subsequent statistical processes.

Nominal

This is the simplest and least sophisticated type of measurement, it involves counting things into categories (or buckets), literally "naming" things. The only value associated with individual members of the population is the value: "1". As the individual members of the population are counted into categories, the subsequent data value is the number per category. With this type of data, how things are alike is that they are in the "same" category, how things are different is that "different" categories exist within the population. There are differences between the categories the population is counted into, but no numerical value or priority to these differences. One category is as meaningful as another. This level of scaling conveys the least information, and all other more-sophisticated levels can be treated as nominal data. Examples of such categories that constitute nominal measures are: male/female; white/black/Chicano/native American; democrat/republican. Nominal data is probably the most common data encountered and is quite commonly used in polls, surveys, and job applications.

One other type of nominal data is any use of numbers as a name or as a unique identifier. A unique serial number, a social security number, a checking account number are all examples of using the (nominal) number itself as a name or identifier. In this instance, even though there may be numerical differences that could be generated between the individual "values" of the numbers, there is no meaning to be ascribed to these differences.

Often when information for the various categories is a large number of counts – like results from a survey or poll – the counts are converted to a percentage or proportion of total counts. For example a pre-election survey showing 279 respondents planning to vote Democrat, 357 Republican, 22 Libertarian and 12 Green Party could be expressed as: 41% (or .41) Democrat, 53% (or .53) Republican, 3% (or .03) Libertarian and 2% (or .02) Green Party. The expression of this nominal data as percentage results does not change the categorical nature of the data and this is still nominal data. (The fact that these percentages add up to 99% instead of 100% is an example of *rounding error*, the fact that when numbers are expressed with a restricted number of digits, some of the data beyond those digits is discarded and may affect subsequent operations like addition, multiplication, etc.)

Ordinal

As implied by the name, this type of data results from the process of rank ordering information from greatest to least. The position of an individual data value within the ordering conveys relative information about priority and relationship to the ranking of the other individual data values. The numerical difference between the value of adjacent rankings is not meaningful other than to convey priority. The value of the difference between first and second (2-1=1) and second and third (3-2=1), does not convey any meaningful information about the true difference between these positions. Examples of ordinal scaling include birth order, class standing, percentile ranking, and order of finish in a marathon. This data can be reduced to nominal data by using the order information to establish categories (i.e. remove the priority information from the data).

A type of ordinal data encountered in surveys or evaluations in the social or applied behavioral sciences is often called a "Likert" scale after Rensis Likert, one of the first behavioral scientists to make extensive use of this kind of measure. In this type of data numbers from 1 to 5 are associated with increasing intensity of responses: 1 = Never, 2 = Sometimes, 3 = Frequently, 4 = Often, 5 = Always. This type of data clearly displays the characteristics of ordinal data: priority among the values but no numerical meaning to the differences between values.

Interval

The name gives this one away. In this type of scaling the differences between the measures (the interval) has defined and consistent units of measure. (Interval is distinguished from Ratio by its lack of absolute zero.) This type of scaling is usually involved whenever we are measuring anything, in a physical sense. Thus if we are measuring temperature the value of 1 degree F is the same whether that 1 degree occurs between 32 and 33 degrees or between 88 and 89 degrees. The consistency of the measurement intervals is the distinguishing characteristic of this level of scaling. Examples are: temperature (degrees) and calendar years - AD or BC. This data can be used to prioritize individual values and thus can be reduced down to either ordinal or nominal data. The distinction between interval and ratio data is important and will be discussed extensively, both types of data are treated the same in all statistical applications.

Ratio

This is the most sophisticated level of scaling and involves data that both has a constant interval value and has an absolute zero. This absolute zero reference means that whenever the quantities being measured have a given ratio the measures themselves will reflect the same ratio. Thus if I weigh 240 pounds and you weigh 120 pounds, I am twice as heavy as you. Interval measures with an arbitrary zero reference will not show the same ratio relationship: a spring day at 66 degrees is not in any sense twice as hot as a winter day at 33 degrees. For all practical purposes almost all physical measures are ratio types of measures. This is the highest, most sophisticated level of scaling. As such, data that is ratio may be reduced to all other levels of scaling.

An Approach to Distinguishing Ratio and Interval Measures[1]:

This approach provides help in distinguishing whether a measure is interval or ratio. If in ordinary, everyday usage the measure is used with negative values (i.e. -12°C), it is most likely an interval measure. Ratio measures with their absolute "zero" will not have negative values.

Comparison of Levels of Scaling

It is critical to understand the type of data you are working with in order to ensure the correct subsequent statistical processes are used. There is a logical flow of questions to ask that will clarify the type of data with which you are working:

1. Does the data have a consistent numerical difference between values (categories)?
   No – continue to question 2. Yes – continue to question 3.
2. Does the data have priority to the sequence of values (categories)?
   No – this is nominal data. Yes – this is ordinal data.
3. Does the data have an absolute zero value? (Are only positive values of the data meaningful?)
   No – this is interval data. Yes – this is ratio data.

---

[1] I am indebted to Bachelor student Marc Whitman for this observation.

Level of Scaling	Characteristics	Examples
Nominal	Categories, no value or priority differences between categories.	Social Security Number, Male/Female, % voting Democrat in Presidential election.
Ordinal	Priority, ordered, no constant value difference between rankings.	Birth order, finish order in a race, percentile ranking in a test, tallest to shortest.
Interval	Constant unit (interval) of measure, no absolute zero value.	Temperature, elevation above or below sea level, calendar year.
Ratio	Constant unit (interval) of measure, absolute zero value.	Most physical measures, weight, length, etc.

Exercise 2.1

Identify the level of scaling (Nominal, Ordinal, Interval or Ratio) for the following italicized measures:

1. __The university student body is *25% male* and *75% female.*
2. __1997 Ford trucks are available in *nine* different body *colors.*
3. __The salmon weighed *12 pounds*!
4. __Cook the casserole at *375 degrees Fahrenheit.*
5. __The salmon won *third* prize at the salmon derby!
6. __It took *47 minutes* to get to work this morning.
7. __The Mariners are *first* in their division.

8. __The same day my wife got a ticket for going *15 mph over the limit* in our city neighborhood, I got a ticket for going *30 mph over the limit* on the freeway.
9. __I was born in *1946*, my son in *1970*, and Julius Caesar was born *101 BC*.
10. __Place *two teaspoons* baking soda in the cake mix.
11. __Being *first-born* gave an advantage to Susan.

If you applied the above 3 questions to distinguish between levels of scaling to the italicized values and data descriptions in exercise 2.1, you probably were able to distinguish the correct answers. Items 1 and 2 are both examples of nominal data. Even though item 1 is expressed as percentages, male and female are clearly categories; as are truck body colors. Items 5, 7 and 11 are ordinal data examples. Whenever data relates solely to priority or order of the data, it is ordinal data. Items 4, 8 and 9 are interval data. Temperature (at least with the Fahrenheit or Celsius scales) always has meaningful negative values and therefore has no absolute zero. Item 8 is trickier; speed would usually be a ratio measure, however, since this is speed relative to 2 speed limits, the speed limits are the arbitrary (not absolute) zero values and these are interval measures. Item 9 is interval since calendar years BC are the equivalent of negative numbers. Items 3, 6 and 10 are ratio measures. The key question used to confirm that they are ratio is: are negative numbers or values to this data meaningful? Since the answer is no, they have absolute zeros. Teaspoons are a physical measure (standard teaspoons are maintained at the former National Bureau of Standards), not just a category.

3 Population Distributions and Measures of Central Tendency

Population Distributions

The first step in describing a population is to define it in such a way that others can re-establish that same population later. The second step is to develop some data about the population that will let us begin the statistical process of simplification. Usually at the time you define the population you will know what you want to measure. It is important at the beginning of your measurement process that you are clear about what type of measure, "level of scaling", you are using to establish the data. This decision made at the beginning will control your ability to perform subsequent statistical operations and interpretations.

So now you've established your population and the data you want to acquire. What is your next step? Go acquire the data. That process could be as complex as the U.S. census measured every ten years, or as simple as asking a few questions of a few people and recording the responses. An important point to remember is that in describing a population you have to have data for the whole population. As you acquire the data (the first step in the simplification process) you will probably realize that there is a spread to the numbers. Whatever characteristic you are measuring about your population will not be the same for every member. It will be spread out in different amounts among the population members.

Visual and Graphic Representation of Distributions

Because of the spread of your population information, you may want to represent that information visually. A visual representation allows us to rapidly assess how the information or data is apportioned among the population members, much more rapidly than looking at a list of numbers. Two common graphical representations that are used are the *histogram* (bar chart) and the *frequency polygon* (line graph). The bar chart/histogram is the visual equivalent of stacking columns alongside each other with the height of each (y-axis) representing the amount of data in the categories or value range (x-axis). The frequency polygon uses a line connecting the tops of what would be the bars in a histogram without displaying the rest of the bars. This line meanders from the baseline value on the x-axis to whatever the largest value is in the population. Examples of each of these two charts follow. The process of preparation of a histogram or frequency polygon is rather common sense. The most important facet is the selection of the intervals across the bottom of the chart. As a general rule of thumb, for ease of interpretation the intervals across the x-axis (bottom of the chart) should be categories (for nominal data) or multiples of 1, 2, 5, 10, etc. for interval or ratio data. This assures that one of the major divisions will be multiples of 10. Because of our ten-based method of counting, this will make a visual presentation most readily interpretable by most people.

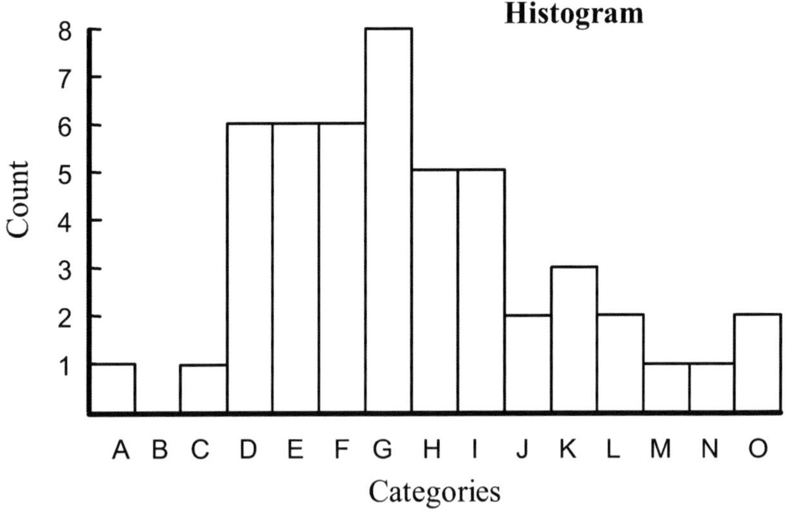

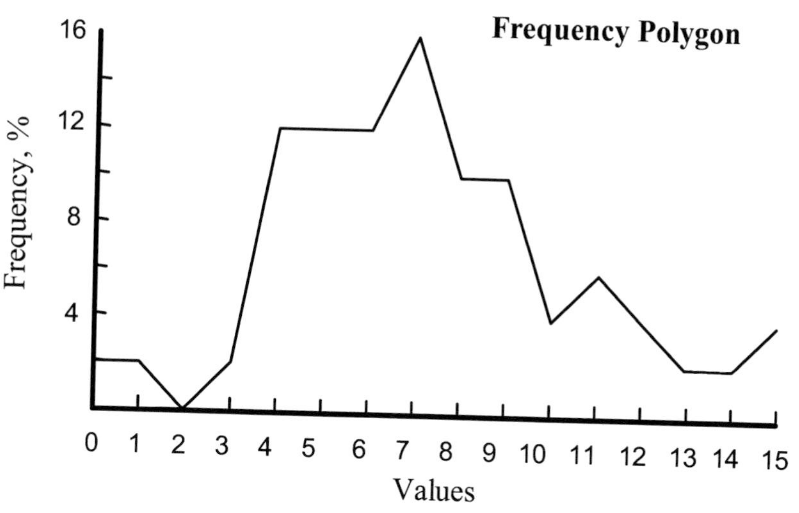

Exercise 3.1

Using the data sets below and the 10 X 10 Grids, create both histograms and frequency polygons to display the data sets. (For nominal data, just create a histogram with each bar indicating the counts in a given category.)

A. A group of 25 students with the following ages:
25, 22, 22, 27, 20, 21, 20, 24, 24, 26, 21, 21, 20, 22, 26, 21, 22, 28, 22, 25, 25, 21, 22, 25, 24

B. Survey data of planned occupations after graduation:
Entrepreneur: 2
Consultant: 7
Counselor: 10
Office/clerical: 0
Manager: 4
Executive: 1
Teacher: 1

A. Histogram

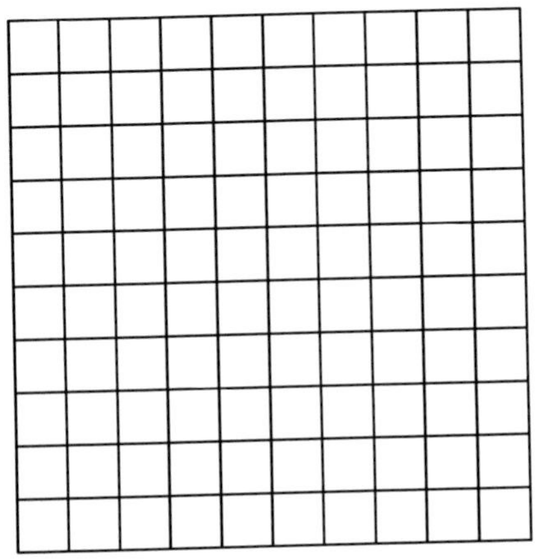

A. Frequency Polygon

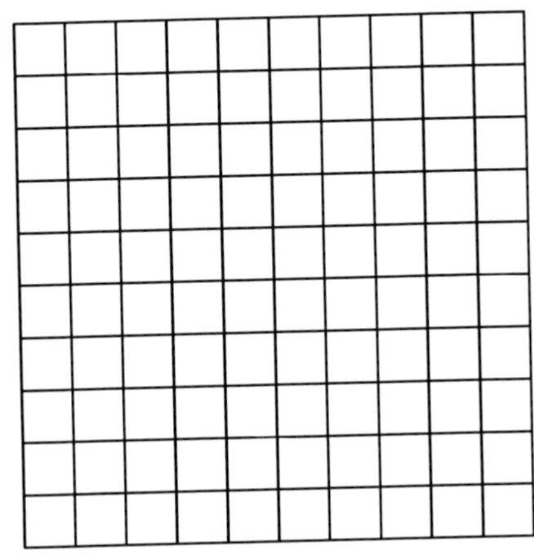

B. Histogram

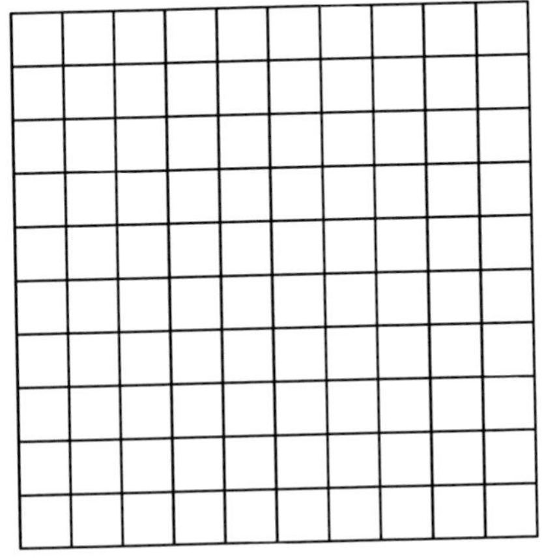

B. Frequency Polygon

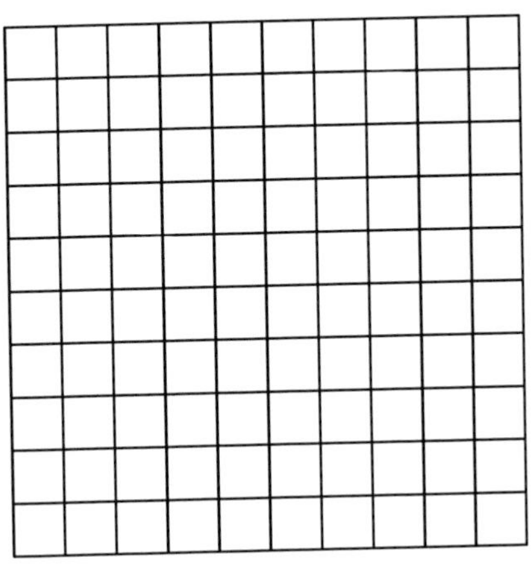

Compare your results with your learning partner. Note that since data set "B" is nominal data (did you catch this?) you do not need to prepare a frequency polygon for it. Histograms/bar charts can be used for all levels of scaling. The frequency polygon/line graph is used usually only for continuous data, which is data that is either interval or ratio.

Examples of two ways to plot these data sets are shown following. There is no single "best" way to determine what range of values constitute the "x" and "y" axes.

A. Example Histogram

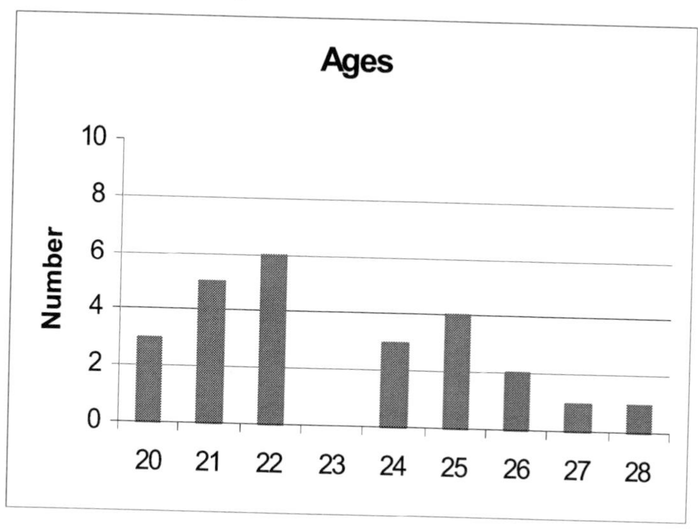

A. Example Frequency Polygon

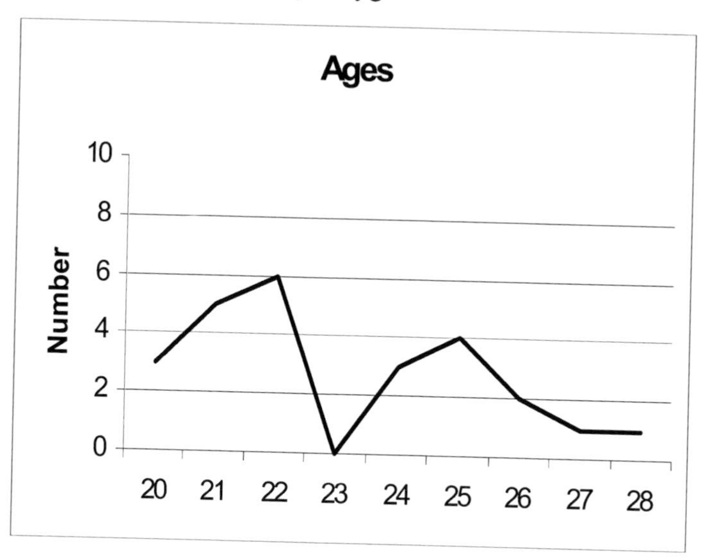

B. Example Histogram (Count Based)

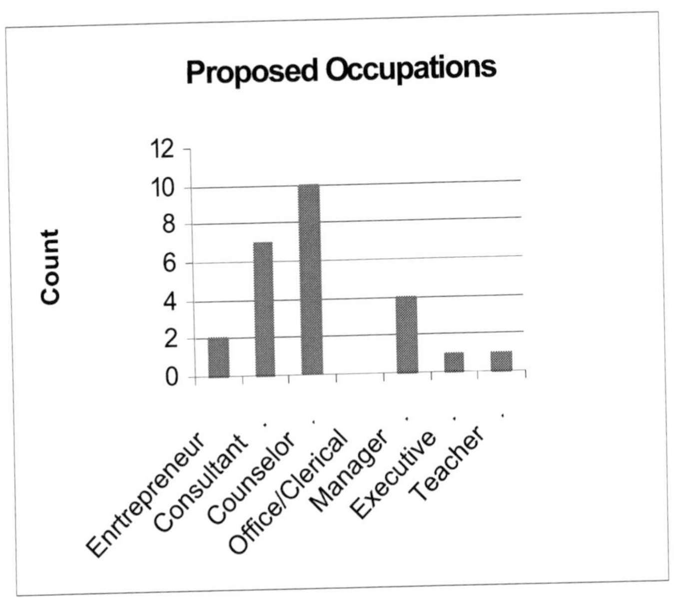

B. Example Histogram (Proportion Based)

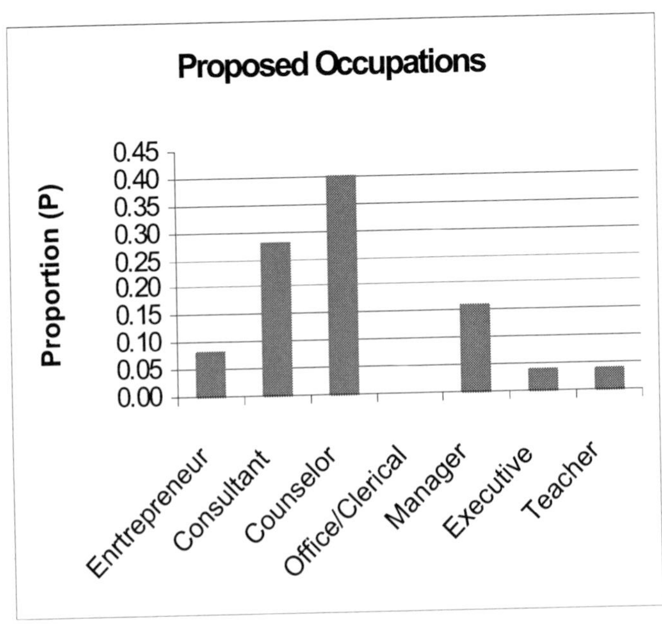

Compare your visual presentations to the four examples above. There are several issues to pay attention to. A value of "zero" in the data is still a value that is included in graphs for both sets of data. Data can be represented as counts/numbers or proportion/percentage of individual values to the whole. In the latter case (the second histogram from Example B), the total of all the proportions of all the categories is equal to 1.0. When distributions are shown as proportion of a complete set, the proportions should always add up to either 1.0 or 100% (except for small differences due to rounding error).

Developing Your Own Population Distribution(s)

It is worthwhile to develop your own sets of "population" data that can be used for practice of the various calculations you will be learning. You can either "dry lab" the data and create values or develop data from your classmates. The size of the data sets should be from 10 to 25 values to ensure that calculations are not trivial and also not overwhelming. Both sets should be built using interval or ratio data. One suggested set is to survey the ages of your fellow students, workers, or others. Another convenient set could be easily obtained from almost any newspaper. Look at the national weather statistics and record high temperatures for a given day from across the nation. Or look at financial pages and record closing stock prices for 10 to 25 stocks for a given day.

It is time to introduce some of the practices or "discipline" that will become increasingly important to your success with statistics. The first practice is useful to keeping track of your data. When we are talking generically about numbers we often use a symbol for the numbers, X. This symbol is called a "variable", meaning the letter "X" can stand in for any of the pieces of data we are working with from your population. In order to keep track of the data we number the "X's" as X_1, X_2, X_3, X_4, etc. up to the final value of data from the given population X_N. (For a population the last piece of data has the number "N" associated with it. Since the last piece of data defines the size of the given population, the value of N for any population is the symbol for its total size and N is synonymous with the *population size*.) Another symbol is the letter "i" which represent the numerical sequence of the number associated with each "X". This *index* symbol will reoccur throughout this book. Thus the generic symbol for any piece of data from any distribution is "x_i" which could be translated into the "i-th value of the data from the given distribution".

If we applied this beginning discipline to the "A" distribution from Exercise 3.1, we would re-format the data as follows (it is formatted in columns to support later calculations that will use the data):

Student Ages from Exercise 3.1A

i	X_i
1	25
2	22
3	22
4	27
5	20
6	21
7	20
8	24
9	24
10	26
11	21
12	21
13	20
14	22
15	26
16	21
17	22
18	28
19	22
20	25
21	25
22	21
23	22
24	25
25	24

Measures Used to Represent the Similarities

After we have completed the preceding steps of simplifying reality, we now have a set of numbers that in some way represent the population that we are dealing with. If we have done the job right we should have as many numbers as there are members of the population. If the population size is three then the data isn't too difficult to make sense of. On the other hand if the population size is 397, then dealing in a sensible way with 397 numbers can be quite difficult. (By the way, the mathematical abbreviation for the number of members of the population as previously described is "N".) The next step then is to find a method to further simplify our "N" numbers down to one number that in some way best represents the center value (the "similar-ness") for the data. There are three such representations that we use: the mode, the median, and the mean (or average). Using the precise language of statistics these three measures of similarities are called *measures of central tendencies*. They tell us about the similarities of the population data. They are also applicable to different levels of scaling and have different strengths and weaknesses as representatives of a population's central value.

Mode

This measure of central tendency is the data value that occurs most frequently in the population distribution or is the nominal category with the greatest number of counts or highest proportion. The mode is usable with nominal data and all higher levels of data (since they may be reduced to nominal data). Thus the mode may actually be determined for all four types of data. The determination of the mode is basically a visual operation that involves looking for the most common value in the data or the peak of a distribution curve. In exercise 3.1A the mode is 22 years old and in exercise 3.1B the mode is the career of Counselor. The mode does not always exist for a population. Some populations have no mode, e.g. no peak or largest category count and are called "amodal". Some distributions will have more than one mode. A population that has two modes is called "bi-modal". The mode is the most restricted measure of central tendency. Its strength is that the mode can be determined (if it exists) for any type of data. The weakness is that it only reflects the most common data value and may be a very poor representation of a center value for the population.

Median

The median is the mid-point of the data when it is arrayed (lined-up in order) from lowest to highest value. Because it requires data that can be ordered, it is applicable only to ordinal and higher level data. To determine the median the data is listed in ascending or descending value, and the value of the data at the exact middle of the ordered data is the median. It is readily determined if the number of data for the population, N, is odd. If, however, N is even, then the median is the mid-point between the two data around the middle. (This is accomplished by adding together the two numbers on either side of the middle and dividing by 2.) The median is most often used as a measure of central tendency when the data at either the upper or lower end of the distribution represent large extremes. For instance data on annual earnings have a lower limit of $0 but no upper limit (some people earn billions $$ annually). Because the median represents the true mid-point of the population it is not affected by extremes at either end. The weakness to the median is that it may be clumsy to determine because the data all has to be manipulated and ordered to determine the median.

For example to find the median from exercise 3.1A, the 25 data points have to be re-ordered from lowest to highest. When so arrayed, the middle (13th) data point in the ordered results is the median. To illustrate: the original data:

25, 22, 22, 27, 20, 21, 20, 24, 24, 26, 21, 21, 20, 22, 26, 21, 22, 28, 22, 25, 25, 21, 22, 25, 24

would be (re)ordered from lowest to highest values. Note that each data point that matches another data value in the original data set is reproduced into the (re)ordered dataset:

20, 20, 20, 21, 21, 21, 21, 21, 22, 22, 22, 22, **22**, 22, 24, 24, 24, 25, 25, 25, 25, 26, 26, 27, 28

As shown, the highlighted value of 22 is at the exact middle of the distribution and therefore is the median. (Note that no median is possible for the exercise 3.1B distribution, which is nominal data and does not contain priority among the categories.)

If we had another data set with an even number of data, the median is calculated as follows, new dataset of 10 ages:

34, 23, 45, 35, 27, 36, 34, 44, 42, 41

(re)ordered data set:

23, 27, 34, 34, 35, 36, 41, 42, 44, 45

The two middlemost values are: **35** and **36**, the true middle (or average) between them is obtained by adding them together and dividing by 2:

$$\frac{35+36}{2}=\frac{71}{2}=35.5$$

which is the median for this distribution.

The median is obviously a "good" representation of the middle of a distribution. It only applies to ordinal and higher level data. It is a particularly valuable measure of central tendency for distributions that are unbounded at one end and which may contain extreme values. Examples of such distributions would be population income data (what is the most money made by any one person?), house sale prices, etc. Because the median only reflects the middle-most value(s) of the population, it is unaffected by extreme values at either side of the distribution. The operational definition of the median is 'the value in the distribution such that half of the distribution data values are less and half the distribution data values are greater". Of course since the median is only reflective of the middle value, it does not reflect how much less or greater are the other population values.

Mean

For the purposes of this study, the mean is the same as the average and the two terms will be used interchangeably. The mean is calculated by adding up all of the data and dividing the total by the number of data points. This is the first measure of central tendency that results from a mathematical operation on all of the population data. (As one student quipped[2], "the mean is inclusive of all diversity") The symbol for the mean of a population is the Greek letter μ pronounced "muu". On the next page is a simple equation defining the mean as the process of adding up all of the data values and then dividing by N. The mean can only be calculated for interval and ratio data because the mean is the balance point where positive and negative differences to the data exactly cancel each other. Thus all the differences between the mean and the rest of the data add up to exactly zero. This is one of the strengths of the mean, that it is a balanced center point that reflects the values of all the data. The fact that it is calculated through a mathematical operation is another strength, because large amounts of data are almost as easy to calculate as small amounts. The weakness is that because it does reflect the values of all the data it is overly affected by large values at either extreme in the population distribution. (If 30 of us are standing around and calculate our average net worth is $100,000.00, then Bill Gates joins us, the average net worth for the 31 of us is $1,000,000,000.00. Yet none of the original 30 is any richer.) As a measure of similarities of the population it is the measure most frequently used.

Calculation of the Mean

The process of calculating the mean is quite simple and mechanical, add up all of the data and divide by the number of data values added together. Statisticians have developed a shorthand method for describing how to perform a mathematical operation such as calculation of the mean. The shorthand equations or formulas say the same thing as the word description above, but much more concisely.

The formula for the mean is:

[2] Insight from masters program student Barbara Grant.

$$\mu = \frac{\sum X_i}{N}$$

...where Σ is the Greek letter "capital SIGMA" and is the mathematical symbol for addition. I like to call it the "add em up" sign because it means just that; add up all the elements to the right of the <u>*summation*</u> or "add em up" sign.

...where X_i is the symbol for the individual data values for the population. As previously described, the letter "X" is universal in mathematics for a variable data value. The "sub-i" is the symbol for the individual data values. The pronunciation of the symbol is "X-sub-i".

...where the line denotes the division operation and the N is the total number of data points in the population.

A simultaneous translation of the equation is: the mean (μ), is calculated by (=), adding up (Σ), all of the individual data values (X_i), and dividing the sum (—), by the number of data points (N).

Exercise 3.2

Calculate the Measures of Central Tendency: mode, median and mean for the following populations: (compare your results with your learning partner)

A. Ages of: 43, 32, 49, 41, 47, 52, 47, 46, 50
mode = _____, median = _____, μ = _____

B. Salmon weights of: 12.0, 5.6, 8.0, 13.4, 16.0, 8.0, 9.0 and 11.5 pounds.
mode = _____, median = _____, μ = _____

Answers to Exercise 3.2A and B

First set the data up in columns:

3.2A		3.2B	
i	X_i	i	X_i
1	43	1	12.0
2	32	2	5.6
3	49	3	8.0
4	41	4	13.4
5	47	5	16.0
6	52	6	8.0
7	47	7	9.0
8	46	8	11.5
9	50		

1. *Determine the modes, the data numerical value which is most repeated:*

for 3.2A inspection shows that the value 47 is the only value repeated, therefore the mode = 47

for 3.2B inspection shows that the value 8.0 is the only value repeated, therefore the mode = 8.0

2. *Determine the medians: first reorder the data from least to highest:*

for 3.2A,
*32, 41, 43, 46, **47**, 47, 49, 50, 52*
(the middle value, #5 out of 9) median = 47

for 3.2B,
*5.6, 8.0, 8.0, **9.0**, **11.5**, 12.0, 13.4, 16.0*
(the middle values, #'s 4 and 5 out of 8 are averaged) median =

$$\frac{9.0+11.5}{2} = \frac{20.5}{2} = 10.25$$

3. *Determine the means:*

for 3.2A,

$$\mu = \frac{43+32+49+41+47+52+47+46+50}{9} = \frac{407}{9} = 45.2$$

for 3.2B,

$$\mu = \frac{12.0+5.6+8.0+13.4+16.0+8.0+9.0+11.5}{8} = \frac{83.5}{8}$$

$\mu = 10.44$

The means could also have been calculated with the column data. Just add up each column and divide by the appropriate value of N. Note that the values of the means calculated above are rounded to one decimal more than the original data. This is an effective rule of thumb for rounding numbers for "final" values of calculations.

4 Measures Used to Represent the Differences.

Measures of Dispersion

As we are developing our simplification of reality, we focus on two measures: one that describes the similarities (Chapter 3) and one that describes the differences. Speaking statistically these numbers that represent the differences are called *measures of dispersion*. If we think about the word "dispersion", it conveys the sense of how the data are spread around the middle. We will consider four measures of dispersion: the range, the average deviation, the variance and the standard deviation.

Because measures of dispersion are based on the values of differences between population measures, data is required that has consistent numerical values for the differences between data. Nominal and ordinal data do not have such consistent measures for differences. Therefore these measures of dispersion only apply to interval and ratio data. (There are methods, different from these, for calculating measures of dispersion for nominal and ordinal data.)

Range

The range is the simplest and also the least informative of measures of dispersion. The range is calculated by subtracting the smallest value in the population from the largest value. The range is usually abbreviated with an "R". An example of calculation of the range: for a population of 397 numbers representing golf scores recorded on a certain weekend, the highest is 129 and the lowest is 54 (for an 18-hole course), the range is

$$129 - 54 = 75.$$

The range is the easiest measure of dispersion to calculate since only one subtraction is required. It may be a cumbersome preparation for the calculation for a large amount of data, however, because the data must be organized or scanned to determine the lowest and highest value. The range is a poor measure of dispersion because it only reflects the extremes of the data and so does not provide any information about the differences among the rest of the population. The fact that the range is based only on two numbers whether the population size is 3, 3000, or 3 million is also very dissatisfying because all the rest of the data is ignored in its calculation.

The range is useful, however, in representing the dispersion of very small amounts of data. If the amount of data, N, is equal to 2 then the range is the only measure of dispersion we can calculate. Likewise for small samples or populations of size 3 or 4, the range may be a very meaningful number for reflecting the dispersion.

(Average Deviation)

It would seem to make sense intuitively that just as the mean may be the most useful measure of central tendency because it reflects all of the data, the best measure of dispersion should also reflect all of the data. There is an inherent dissatisfaction with using the range, since it only reflects the data at the population extremes. It would be more satisfying to develop a measure of dispersion that reflects the differences of the population data not from the extremes, but from the center value. One approach would be to determine the differences or deviations of the data from the mean as the center value (since the mean is calculated from all the population data). The average of these differences would be a single number that would represent these differences.

Let's see if we can develop an operational definition for calculating an average difference or "deviation" for a given set of data. What we would want to do is subtract each individual data value from the mean value, add up all the differences, and divide by N to get the "average deviation." We could prepare a mathematical shorthand definition of the average deviation similar to that presented for the mean in the last chapter. The first step is to subtract each data value (X_i) from the mean (μ):

$$(\mu - X_i)$$

Then we need to add together all the differences;

$\Sigma\,(\mu - X_i)$

Finally to find the average value of these differences we divide the total differences by the number of data:

$$\text{average deviation} = \frac{\sum(\mu - X_i)}{N}.$$

Let's try an example to see how this measure of dispersion works. Assume that we have a population of students with the following ages:

40, 24, 37, 43, 36, 30.

Let's set up a series of columns to allow us to perform the required operations on these numbers to calculate the mean and average deviation. The columns are set up below, note that the first column has the index (i) which is set up to make keeping track of the data easier.

i	X_i	$\mu - X_i$	$\mu - X_i$
1	40	35 - 40	-5
2	24	35 – 24	11
3	37	35 – 37	-2
4	43	35 – 43	-8
5	36	35 – 36	-1
6	30	35 – 30	5
	Σ 210		Σ 0

Notice that in order to prepare the third column to calculate the differences we must first add up the second column and divide by six to establish the mean,

$$\mu = \frac{210}{6} = 35.$$

Having prepared the fourth column, a list of the individual deviations, we now add them up to get the total deviations and divide by 6 to get the average deviation. But what happens when we add the column? We find that the total of the deviations is zero. That is all of the positive deviations exactly cancel all of the negative deviations. Is this result unique to this particular sample? No it is not! If you will recall we discussed the fact that one of the properties of the mean is that it is a balance point for a population where all the positive differences just equal all the negative differences. Thus the average deviation will always be zero since the sum of the positive deviations always equals the numerical value of the sum of the negative deviations. Equal positive and negative numbers always add to zero. This fact makes the average deviation (which seemed like a good idea at the time) useless as a measure of dispersion. As we work to find another way to determine the differences, we will continue to calculate average deviation for a short period to drive home the reality of its non-usability.

It is possible to compensate for the balancing effect by ignoring the negative signs and making all deviations positive. This is called taking the absolute value of the deviations. This process will give us a number that we can use to represent the average deviation, however, it is inherently dissatisfying to ignore the information contained in the sign (+ or -) of the deviation. It is equivalent to saying the 43-year-old and the 24-year-old in our example population are essentially the same age. In addition an average deviation calculated using absolute value would not be usable in other statistical applications. For this reason we calculate another measure of dispersion: the standard deviation.

Variance and Standard Deviation:

When mathematicians are faced with a problem of negative values counterbalancing positive values, they can get very clever to legitimately remove the effect of the negative sign. One very effective way to remove the negative sign from a number is to multiply it by itself (square it) since two negative numbers multiplied together always produce a positive product. Because the product of multiplying two positive numbers is also positive, the squaring operation always results in a positive product. Thus we would have a useable measure of deviations if before we add up all the deviations we square them. Thus instead of the average deviation we would get the average squared deviation. For the population data previously used we have to create another fourth column that is the squared deviations. The process would then look like:

i	X_i	$\mu - X_i$	$(\mu - X_i)^2$
1	40	-5	25
2	24	11	121
3	37	-2	4
4	43	-8	64
5	36	-1	1
6	30	5	25
	Σ 210	Σ 0	Σ 240

In this case when we add up the fourth column, the squared deviations, we do get a number other than zero. This particular sum is 240. When we divide that sum by six we get the "average squared deviation" that is called the *variance* and is equal to 40. Thus we have very cleverly taken a situation in which we were unable to get any result other than zero, and by adding one step (squaring) we are able to get a (nonzero) measure of dispersion. But maybe we are not so clever, because what use is it to know that the (average squared deviation) variance is 40? We deal with ages (years), not squared ages (years^2)! Perhaps we need to add one more step, we need to determine the square root of the variance. That square root is called the *standard deviation.* (The square root is the positive number which when multiplied by itself equals the number we are finding the square root of.) The symbol of the standard deviation of a population is σ, the Greek letter "sigma" - in the lower case. For the previous sample where the variance (average squared deviation) was 40, the square root of 40 is approximately 6.3, which we call the standard deviation of this population.

The formula for the variance and the standard deviation reflects the process we have just developed:

$$\text{variance} = \frac{\sum(\mu - X_i)^2}{N}, \text{ and } \sigma = \sqrt{\frac{\sum(\mu - X_i)^2}{N}}$$

The priority of the operations is as follows: first, calculate the mean; second, subtract the individual data values from the mean; third, square each of the differences; fourth, add up all of the squared differences; fifth, divide by the number of data values (this result is the variance);

sixth, take the square root of the result of the division (this result is the standard deviation). Thus the variance and standard deviation have a square and square root relationship with each other:

$$\sigma = \sqrt{\text{variance}} \quad \text{and} \quad \text{variance} = \sigma^2$$

The variance is an important measure of dispersion in statistics, and is used frequently in sophisticated statistical analyses where the dispersions of different populations are compared. However, since the variance is in "squared units", we find the standard deviation much more useful because it is expressed in the same units as the original population values.

The standard deviation is a very useful measure of dispersion. The standard deviation includes all of the data from the population in its calculation, so it is a comprehensive measure. The standard deviation employs a mathematically rigorous method to eliminate the effects of the negative signs. Finally because of the extreme usefulness of the probability relationships that can be developed using the standard deviation and the normal curve, it is universally accepted as the most useful measure of dispersion.

Before performing calculations of standard deviations and demonstrating an easier formula for making the calculation, I would like to comment that the mean and standard deviation are the heart of both descriptive and inferential statistics. In our analogy to verbal descriptions of groups, we pointed out that you can only describe effectively by noting similarities and differences. The same is true in statistics, you must have two numbers; one which measures similarities (the mean) and one which measures differences (the standard deviation). With these two numbers that are a great simplification of the data (reality) we can not only describe chunks of reality (populations) but we can also use the mean and standard deviation to make predictions about populations based on data from a sample of the population.

Exercise 4.1
Calculate the following values for the three populations in the tables below: the mean, the average deviation, the variance (average squared deviation), and the standard deviation (square root of the variance):

Population A:

index	X_i	$(\mu - X_i)$	$(\mu - X_i)^2$
1	15	______	______
2	15	______	______
3	15	______	______
4	15	______	______
5	15	______	______
6	15	______	______
7	15	______	______
8	15	______	______
9	15	______	______
10	15	______	______

$\Sigma =$ ______ $\Sigma =$ ______

$\mu =$ ______, Average Deviation $= \dfrac{\sum(\mu - X_i)}{N} =$ ______,

Variance $= \dfrac{\sum(\mu - X_i)^2}{N} =$ ______, $\sigma = \sqrt{\dfrac{\sum(\mu - X_i)^2}{N}} =$ ______

Population B:

index	X_i	$(\mu - X_i)$	$(\mu - X_i)^2$
1	15	______	______
2	13	______	______
3	17	______	______
4	15	______	______
5	14	______	______
6	16	______	______
7	15	______	______
8	16	______	______
9	14	______	______
10	15	______	______

$\Sigma =$ ______ $\Sigma =$ ______

$\mu =$ ______, Average Deviation $= \dfrac{\sum(\mu - X_i)}{N} =$ ______,

$$\text{Variance} = \frac{\sum(\mu - X_i)^2}{N} = \_\_\_\_\_\_, \quad \sigma = \sqrt{\frac{\sum(\mu - X_i)^2}{N}} = \_\_\_\_\_\_$$

Population C:

index	X_i	$(\mu - X_i)$	$(\mu - X_i)^2$
1	15	______	______
2	20	______	______
3	10	______	______
4	15	______	______
5	25	______	______
6	5	______	______
7	15	______	______
8	13	______	______
9	17	______	______
10	15	______	______
		$\Sigma =$ ______	$\Sigma =$ ______

$$\mu = \_\_\_\_\_\_, \quad \text{Average Deviation} = \frac{\sum(\mu - X_i)}{N} = \_\_\_\_\_\_,$$

$$\text{Variance} = \frac{\sum(\mu - X_i)^2}{N} = \_\_\_\_\_\_, \quad \sigma = \sqrt{\frac{\sum(\mu - X_i)^2}{N}} = \_\_\_\_\_\_$$

As you review the results of these calculations, let's notice some of the characteristics of the results. First notice for all three populations that the average deviation for each is exactly equal to zero. We will now dispense with any need to calculate or think about average deviation. (Our previous use of average deviation was a learning aid – to help create understanding of the necessity of squaring the values of differences.) Let's focus on the measures of dispersion that work. For Population A, notice all ten values of the data are 15, so the mean is equal to 15, and all differences from the mean are zero. When no variation or differences exist in data, the measures of dispersion are also equal to zero, as are the variance and standard deviation for Population A.

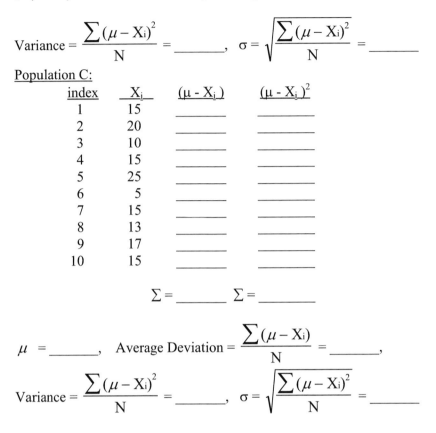

In population B there is a small amount of variation around the mean of 15. The amount of variation is 1 to 2 units. When we look at the measures of dispersion, we see that the range is 4, the variance is 1.2, and the standard deviation (1.1) is just over 1 unit. One way to create meaning for this standard deviation is to add the value of the standard deviation to the mean value and also subtract the standard deviation from the mean. This creates a range around the mean that encompasses <u>*most*</u> *of the actual data values of the population. Perform this calculation for population B:*

$\mu + \sigma =$ ______ *and* $\mu - \sigma =$ ______

Now scan population B and count how many of the population fall within the range of the two numbers above. The majority of the data will fall within this range (more about this later).

In population C there is a much larger variation around the mean of 15. The amount of variation is 5 to 10 units. When we look at the measures of dispersion, we see that the range is 20, the variance is 25.8, and the standard deviation (5.1) is just over 5 units. If we were again to add the value of the standard deviation to the mean value and also subtract the standard deviation from the mean we again create a range around the mean that encompasses most of the actual data values of the population. Perform this calculation again for population C:

$\mu + \sigma =$ ______ *and* $\mu - \sigma =$ ______

Again scan population C and count how many of the population fall within the range of the two numbers above. Again the majority of the data will fall within this range.

You should be developing a sense of the meaning of the standard deviation as a reflection of variation in the population distribution relative to the mean. The majority of the data in a population will always be within one standard deviation on either side of the mean. Another way of saying this is the standard deviation's value reflects how tightly the population data "clumps" around the mean value.

An Easier Method to Calculate Variances and Standard Deviations

The calculation of the variance and standard deviation using the method previously described is tedious and difficult particularly for large amounts of data. There are equivalent calculations that are much easier to use because no subtractions are performed on the individual data that still produce the same results as the longer methods. The equations for this easier method are:

$$\text{Variance} = \frac{\Sigma X_i^{\,2} - N\mu^2}{N} \qquad \sigma = \sqrt{\frac{\sum X_i^2 - N\mu^2}{N}}$$

Perform these calculations as follows: first, square the individual data values and then add them all together; second, from that sum subtract the square of the mean times the number of data (N); third, divide the difference by the number of data and fourth, take the square root to get the standard deviation. This process is an improvement over the other calculation because no individual differences are calculated and the data for each individual are handled only once. Let's apply this equation using the previous data:

i	X_i	X_i^2
1	40	1600
2	24	576
3	37	1369
4	43	1849
5	36	1296
6	30	900
Σ	210	7590

With this data we perform the following calculations:

$\mu = 210/6 = 35$

$\Sigma X_i^2 = 7590$

$N\mu^2 = (6)(35)^2 = (6)(1225) = 7350$

$$\sigma = \sqrt{\frac{7590 - 7350}{6}} = \sqrt{\frac{240}{6}} = \sqrt{40} = 6.3$$

Note that the intermediate product just before we take the square root to get σ, is the variance.

Thus this shortened and easier method to calculate the standard deviation does produce the exact same results as the more difficult method. It does not, however, intuitively convey the definition of the standard deviation as the method employing deviations from the mean does. Because the inherent understanding of the reasons and meanings of the standard deviation are as important as the ability to calculate, I recommend that you perform a large number of calculations of standard deviations using both methods in parallel. Then, if you prefer, use the shorter calculation because of its convenience.

Exercise 4.2

Calculate the Measures of Dispersion: Range, Variance and Standard Deviation for these populations. Apply the disciplines of creating the columns for your calculations. Use both methods for calculating variance and standard deviation: deviations from the mean and the alternate equation.

A. Ages of: 43, 32, 49, 41, 47, 52, 47, 46, 50
Range = _____, Variance = _____, σ = _____

B. Salmon weights of: 12.0, 5.6, 8.0, 13.4, 16.0, 8.0, 9.0 and 11.5 pounds.
Range = _____, Variance = _____, σ = _____

Check your results against these:

4.2A Ages of: 43, 32, 49, 41, 47, 52, 47, 46, 50

1. By inspection we notice the highest value = 52, the lowest = 32.
Range $= X_H - X_L = 52 - 32 = 20$

2. Variance and σ *by both methods:*

a.

i	X_i	$(\mu - X_i)$	$(\mu - X_i)^2$
1	43	2.2	4.84
2	32	13.2	174.24
3	49	-3.8	14.44
4	41	4.2	17.64
5	47	-1.8	3.24
6	52	-6.8	46.24
7	47	-1.8	3.24
8	46	-0.8	0.64
9	50	-4.8	23.04
	Σ 407		Σ 287.56

b.

i	X_i	X_i^2
1	43	1849
2	32	1024
3	49	2401
4	41	1681
5	47	2209
6	52	2704
7	47	2209
8	46	2116
9	50	2500
	Σ 407	Σ 18693

for both methods: $\mu = \frac{407}{9}, = 45.2$

a. variance $= \frac{\sum (\mu - X_i)^2}{N}$ $\quad \sigma = \sqrt{\frac{\sum (\mu - X_i)^2}{N}}$

a. variance $= \frac{287.56}{9}$ $\quad \sigma = \sqrt{\frac{287.56}{9}}$

a. variance $= 31.95$ $\quad \sigma = 5.7$

b. variance $= \frac{\sum X_i^2 - N\mu^2}{N}$ $\quad \sigma = \sqrt{\frac{\sum X_i^2 - N\mu^2}{N}}$

b. variance $= \frac{18693 - 9(45.2)^2}{9}$ $\quad \sigma = \sqrt{\frac{18693 - 9(45.2)^2}{9}}$

$$b.\ variance = \frac{18693 - 9(2043.04)}{9} = \frac{18693 - 18387.36}{9}$$

$$= \frac{305.64}{9} = 33.96 \qquad \sigma = \sqrt{33.96} = 5.8$$

4.2B Salmon weights of: 12.0, 5.6, 8.0, 13.4, 16.0, 8.0, 9.0 and 11.5 pounds.

1. By inspection we notice the highest value = 16.0, the lowest = 5.6.
Range = $X_H - X_L = 16.0 - 5.6 = 10.4$

2. Variance and σ *by both methods:*

a. i	X_i	$(\mu - X_i)$	$(\mu - X_i)^2$	b. i	X_i	X_i^2
1	12.0	-1.56	2.43	1	12.0	144
2	5.6	4.84	23.43	2	5.6	31.36
3	8.0	2.44	5.95	3	8.0	64
4	13.4	-2.96	8.76	4	13.4	179.56
5	16.0	-5.56	30.91	5	16.0	256
6	8.0	2.44	5.95	6	8.0	64
7	9.0	1.44	2.07	7	9.0	81
8	11.5	-1.06	1.12	8	11.5	132.25
	Σ 83.5		Σ 80.62		Σ 83.5	Σ 952.17

for both methods: $\mu = \frac{83.5}{8}, = 10.44$

$$a.\ variance = \frac{\sum(\mu - X_i)^2}{N} \qquad \sigma = \sqrt{\frac{\sum(\mu - X_i)^2}{N}}$$

$$a.\ variance = \frac{80.62}{8} \qquad \sigma = \sqrt{\frac{80.62}{8}}$$

$$a.\ variance = 10.08 \qquad \sigma = 3.17$$

$$b.\ variance = \frac{\sum X_i^2 - N\mu^2}{N} \qquad \sigma = \sqrt{\frac{\sum X_i^2 - N\mu^2}{N}}$$

$$b.\ variance = \frac{952.17 - 8(10.44)^2}{8} \qquad \sigma = \sqrt{\frac{952.17 - 8(10.44)^2}{8}}$$

$$b.\ variance = \frac{952.17 - 8(108.99)}{8} = \frac{952.17 - 871.95}{8} = \frac{80.22}{8} = 10.03$$

$$\sigma = \sqrt{10.03} = 3.17$$

There are several opportunities to learn more about these calculations and how real data affects the final values. Probably the first thing that you noticed is that the resultant values of variance and standard deviation for the different calculation methods were not necessarily exactly equal.

Example

	Calculated by a.		Calculated by b.	
	variance	σ	*variance*	σ
4.2A	*31.95*	*5.7*	*33.96*	*5.8*
4.2B	*10.08*	*3.17*	*10.03*	*3.17*

The differences are not indicative of mistakes or miscalculations. They reflect rounding errors that occur throughout the calculations. The more the data from intermediate calculations is rounded, the larger will be the accumulation of rounding error. Since the calculations in method b involve less intermediate calculations, the rounding error is less.

Another issue to understand is the issue (again) of how much to round the answers to these calculations. Again the scientific rule of thumb is to round the final answer to one digit more that the original data. What about the intermediate results of calculations? Best practice is not to round at all, just carry all the information your calculator is producing in the intermediate calculations. This approach works great if your calculator is sufficiently advanced to be able to complete a series of calculations without having to write down intermediate results. If you have to write down intermediate calculations, then the next best practice is to carry at least one more digit that the final answer will have. This was the practice followed in the calculations above. If all intermediate data were rounded to the same number of digits as the final answer, the differences in calculation results in the table above would have been much greater.

This nearly completes the discussion of descriptive statistics. The next chapter provides information about the normal curve and how its properties add numerical precision to the statement "The majority of the data in a population will always be within one standard deviation on either side of the mean". With the precision provided by the properties of the normal curve, population values of the mean and standard deviation can be used to make probability predictions, evaluate others' statistical statements, and develop statistical tests.

To summarize key aspects of descriptive statistics:

- Descriptive statistics apply to measures developed for a total population.
- In general, Greek letters or capital Roman letters are used for symbols for population values: μ, σ, N, P.
- The collective term for measures and values that apply to populations is <u>*parameters*</u>. An example of the usage of this term is: Values of μ, σ, N and P are parameters for a given population.

<u>Validity and Reliability</u>

These two concepts relate to measurements and have particular application to those measures performed in the behavioral sciences. Reliability is concerned with the repetitiveness or reproducibility of measurements and thus is most related to the standard deviation. Example issues of concern are: can the same results be obtained on different days or with changed (presumably unrelated) environmental factors? The most common issue of reliability is "test-retest reliability" - does a test produce essentially the same results when reapplied to the same person at a different time? Applied to technical measurements, the issue of reliability translates to the precision of the measurements. The illustration below of bull's-eyes shows examples of low and high reliability applied to a pattern of arrows in a target.

Validity is concerned with the issue of whether you are indeed measuring what you claim to measure. This issue is related to the value of the mean and whether it is a true value. This question is very important in psychological and sociological areas because tests or measures may be affected by extraneous and uncontrolled factors. For instance does an IQ test really measure intelligence or does it measure ability to take tests? Examples of validity related areas are: *content validity*, is the result of the measurement really reflecting the desired issue; *face validity*, is the measurement credible on its surface, does it look like a valid measurement (is a typing test appropriate for a person applying for a sales job); *criterion-related validity*, is the measure an effective predictor of future behaviors or other criteria (for example, does a pre-employment test really predict ability to perform a given job)? For technical measurements the equivalent issue is the accuracy of the measurement. The bull's-eyes following show examples of high and low validity.

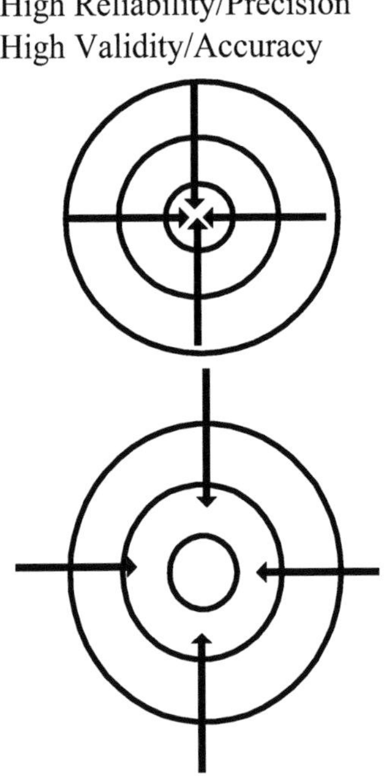

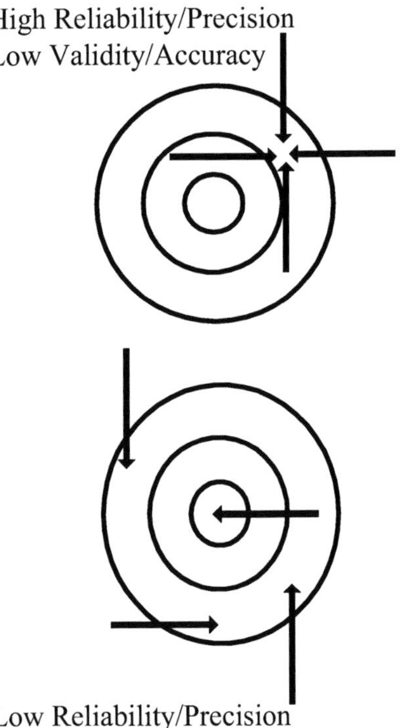

Exercise 4.3

Take a moment and reflect on your understanding of statistics now. Write a few comments on the space below. What does this say about your progress in learning statistics?

____

____

____

____

____

____

5 Wrapping up Descriptive Statistics

Probability Concepts

When in Chapter 2 we drew frequency polygons or histograms to represent the distributions of populations, we plotted the frequency or total number of occurrences of the observation on the "y" axis (vertical scale). Such distributions would be more useful if they were not dependent on the sample size, N, of the population or distribution. One way to remove the specific data on sample or population size is to plot the "y" axis as a proportion. (For a population the symbol for a proportion is p.) Thus if out of a total of eight students, two ages fell in the 30-34 range, the data could be plotted as 2 observations or as a proportion of 2 out of 8 or 2/8 or p = .25. The advantage of plotting distributions as proportions is that the resulting representation is insensitive to sample or population size.

When we convert a frequency to a proportion, we are also generating a number reflecting the *probability* of the observation. The probability of an event or observation is the number of times that event or observation would occur among all events or observations that could occur. The probability is expressed as a decimal fraction, with the probability for any event having a value between zero (the event will never occur) and one (the event will always occur). (Probability is also often expressed as the percentage equivalent to the decimal fraction. Expressed as a percentage, probabilities are a value between 0% and 100%.) The concept of probability is very important to descriptive statistics, because if we have descriptive information about the population we can ascribe a probability to an event or events showing up in a sample from the population. Likewise the concept of probability is absolutely vital to inferential statistics in which we make predictions (probability statements) about the population based on statistics from the sample. The most useful relationship between probability and the mean and standard deviation is the theoretical relationship called the normal distribution.

Exercise 5.1

Several common events have well known probabilities associated with their occurrences. For example when tossing a coin, the probability of heads is 0.5 (assuming a "fair" coin), and the probability of tails is 0.5, providing a total probability of 1.0.

1. Take a coin and record the proportion of heads and tails achieved by multiple tosses. What is the proportion after 10 tosses?_____/_____, after 20 tosses? _____/_____, after 50 tosses? _____/_____, after 100 tosses? _____/_____.

2. Think about a "die", half of a pair of dice, with six faces with values from 1 to 6. Calculate the probabilities of the following events with a "die": "1" _____, "2" _____, "3" _____, "4" _____, "5" _____, "6" _____, Total Probability (add all six values) _______.

3. Now think about two of these dice and the combinations that can be created from the combined values of both. Notice that many combinations will produce some of the middle values, for instance to achieve a "4" value, the following combinations can occur: 1 + 3, 2 + 2, 3 + 1. How many total combinations can be created for the two dice with six faces each? If we consider one die on the left and one on the right, for each unique value showing on the left, six possible values can show up on the right. Thus for the six possible unique values on the left, there are six possible values for each on the right. The total is 6 x 6 = 36. Thus for the value "4", there are 3 out of 36 possible combinations of faces that produce this result leading to a probability of 3/36 = 1/12 = .083.

Calculate the probabilities associated with the following values from two dice:

"2" = __________
"3" = __________
"4" = ___.083___
"5" = __________
"6" = __________
"7" = __________
"8" = __________
"9" = __________
"10" = __________

"11" = ________
"12" = ________

4. (Ideal activity to do with your learning partner). Get a pair of dice and try out how well the above probabilities fit an experiment. Prepare a similar list above for the values that can show up. Roll the dice a hundred times and record a "tick" mark alongside the value each time that value is obtained from a roll of the dice. For a hundred rolls, what proportions of results do you achieve?

1. *The proportion of heads and tails you achieved for the various numbers of coin tosses should approximate 50/50, particularly as the number of tosses gets very large.*

2. *Your calculations for item 2, the six faces of a "die" should all be individual probabilities of .167. The total probability should equal 1.0 but due to rounding error may be slightly larger or smaller than 1.0. In this case 6 x .167 = 1.002.*

3. *Calculations for item 3 require keeping track of the unique combinations that will produce the given result. Easiest way to keep track is to prepare a table or columns:*

Outcome for Both Dice	Left Die Face	Right Die Face	Number of Dice Combinations	Probability of Outcome
2	1	1	1	.028
3	1	2		
	2	1	2	.055
4	1	3		
	2	2		
	3	1	3	.083
5	1	4		
	2	3		
	3	2		
	4	1	4	.111
6	1	5		
	2	4		
	3	3		
	4	2		
	5	1	5	.139
7	6	1		
	5	2		
	4	3		
	3	4		
	2	5		
	1	6	6	.167
8	6	2		
	5	3		
	4	4		
	3	5		
	2	6	5	.139
9	6	3		
	5	4		
	4	5		
	3	6	4	.111
10	6	4		
	5	5		
	4	6	3	.083
11	6	5		
	5	6	2	.055
12	6	6	1	.028
		Σ	36	.999

Note the increasing probability for outcome for both dice progressing from 2 to 7, then decreasing from 7 to 12. The combined probabilities of all outcomes should add up to 1.00. The value of .999 reflects the effect of rounding error for the individual probabilities calculated. Looking at the probabilities of the various outcomes, can you understand that in the game of "craps" the number seven is a loser for the player and a winner for the house?

The Normal Distribution

The *normal distribution* is a mathematical relationship that relates a probability distribution to the mean and standard deviation for a population. It is an artificial, totally unreal relationship defined from an equation, which does do a very good job of representing reality most of the time. The normal distribution is most often represented as a curve, like that below. The curve is either called the normal curve or often the standard curve. The area contained between the curve and the baseline represents probability and totals 1.00 or 100% for the total curve. Before discussing how to understand and apply this curve it is important to be aware of the properties of the curve as defined by its mathematical equation.

The normal curve is *symmetrical*, that is the right-hand side is a mirror image of the left-hand side. Therefore the centerline exactly divides the curve in half, a definition of the median. Since the curve peaks at the middle, the centerline is also the mode for the distribution. Since the curve is symmetrical around the middle, the left side exactly mirrors the right side, the centerline is the "balance point" for the curve and is therefore the mean. So this one property of symmetry combined with the shape of the curve dictate that for the normal curve the mean, median and mode are exactly equal.

The mathematics of the normal distribution define the shape of the curve as a function of μ and σ. The mean, μ, is the center of the curve. The values to the right of the center are marked at $\mu+1\sigma$, $\mu+2\sigma$, and $\mu+3\sigma$. The values to the left of the center are marked at $\mu -1\sigma$, $\mu -2\sigma$, and $\mu -3\sigma$. Along the bottom of the curve are arrows relating the probability with the corresponding area of the curve. The area is the space in the curve between the top line of the curve, the straight bottom line and between the respective markers for $\mu + n\ \sigma$ and $\mu - n\ \sigma$. This probability relationship is fundamental to inferential statistics. The area contained within one standard deviation of the mean (from -1σ to $+1\sigma$) is approximately 68**%** of the area of the curve. Another way of saying this is that for a normally distributed population approximately 68% of the events, observations, or population members will have values within ±1 (this symbol means "plus or minus") standard deviation of the population mean. Thus the probability of any population member having a value within ±1 standard deviation of the mean is 0.68 or 68%. The corresponding area and probability for ±2 standard deviations is approximately 95**%** (or 0.95) and for ±3 standard deviations is approximately 99.75**%** (or 0.9975).

Normal Curve

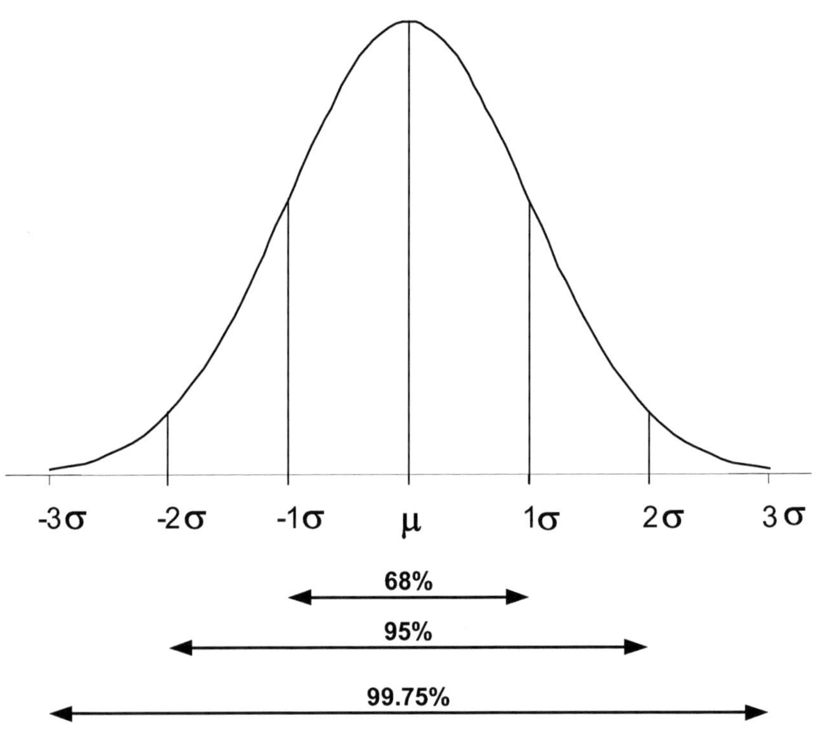

In using the normal distribution to describe a population, the probability as a function of standard deviations can be used to develop probability statements. Thus for <u>any</u> normally distributed population the probability of any single value for the population being between $\pm 1\sigma$ of the mean is 0.68 or 68%. What is the probability that any value of the population will be outside the $\pm 1\sigma$ range? The answer is 1.00 minus the probability it will be within the range, or 1.00-0.68 = 0.32. If we wanted to speak about proportions, 32% of the population have values outside the $\pm 1\sigma$ range.

The normal curve as it extends to either side of the mean approaches the baseline, but it never touches the baseline. In the language of mathematics, the curve is said to "asymptotically approach the baseline". The way this works is: as the curve looks like it is nearing the line, magnifying the curve shows it has approximately the same shape as the unmagnified curve. This is illustrated on the following page. The curve from 3σ to 4σ is compared to a magnified curve from 4σ to 5σ, which in turn is compared to a magnified curve from 5σ to 6σ. So What? The normal curve is not "closed", it extends to infinity in both directions and thus has extremely small values of probability associated with very large distances from the mean. In a practical sense, what does this mean? In most circumstances we will deal with probability values like 90 to 95% or in extreme cases 99%. The corresponding "remaining probabilities" of 10 to 5% or the extreme case of 1% may seem like reasonably small numbers. Who would ever deal with the probabilities associated with distances like 6σ from the mean? How about companies like Motorola, Allied-Signal and General Electric? All three have heavily engaged in improving the quality of their products to "six-sigma" levels. At that level (assuming the six-sigma defines the probability associated with product that does not meet quality standards) what is the probability? It is approximately 4 parts per million or 0.000004. The corresponding probability of products meeting quality standards at six-sigma is 99.9996% or 0.999996. Thus extreme values of the normal curve can be very meaningful.

The normal curve can be used to calculate probabilities associated with specific segments of the curve by combining the known probability values with the curve's symmetry. Because the curve is symmetrical and

The Curve Approaches but Never Touches the Line

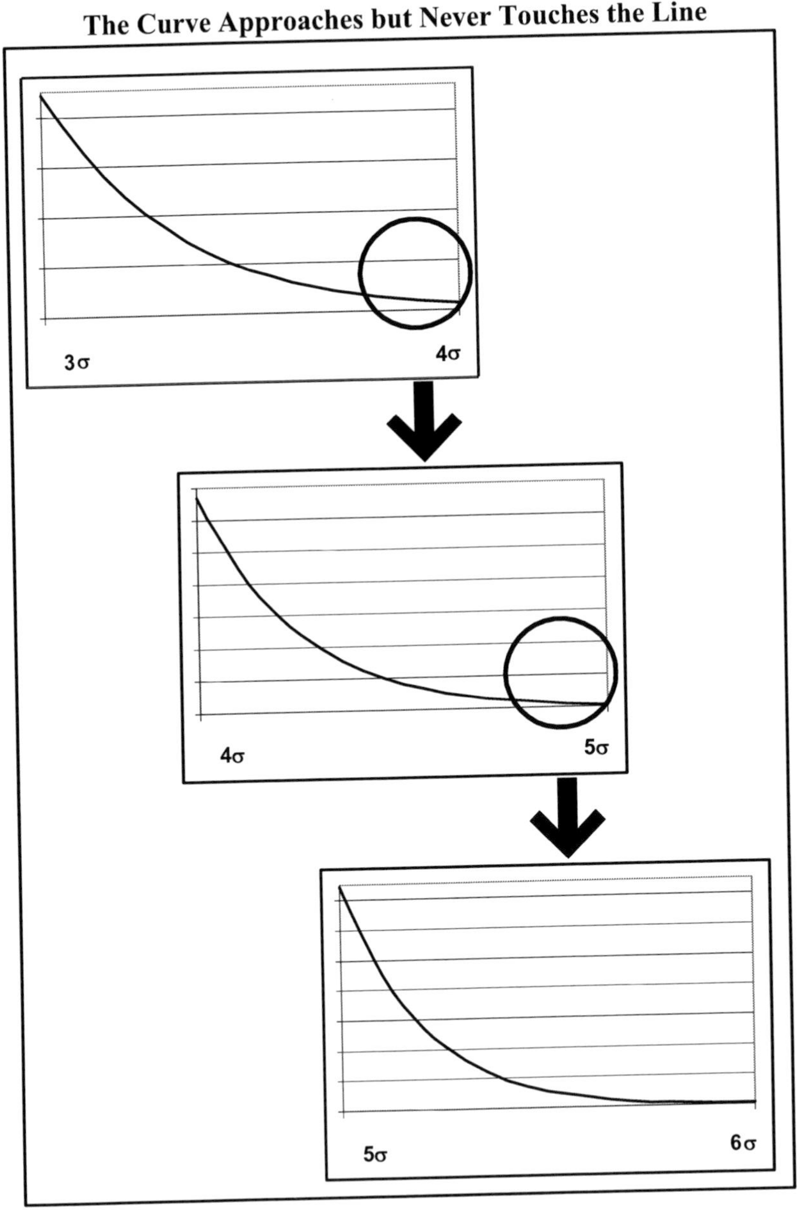

the median and mean are equal, either half of the curve contains a proportion or probability of 0.5 or 50%. How would we find the probability that a member of the population would have a value greater than 1σ above the mean? We would calculate as follows:

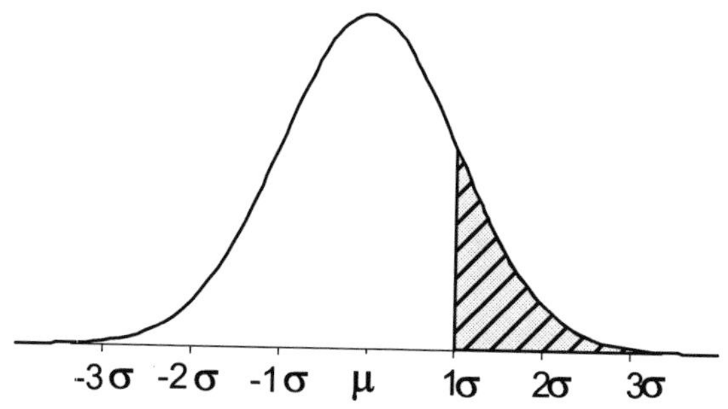

- the probability associated with the left half of the curve is 0.5
- the probability associated with μ to +1σ is one-half of 0.68 or 0.34
- the total probability from the left end of the curve up to 1σ (the total unshaded area) is therefore 0.5 + 0.34 = 0.84
- the probability in the area beyond +1σ (the shaded area) is therefore 1.0 - 0.84 = 0.16.

With practice it becomes pretty easy to evaluate probabilities for exact multiples of σ just by using the symmetry properties of the normal curve and the probability relationships of 68%, 95% and 99.75%.

Exercise 5.2
Calculate the probabilities associated with the following segments of the normal curve using the known values of 68%, 95% and 99.75% and the normal curve's symmetry.

1. *What is the probability associated with the area from* -2σ *to* $+1\sigma$ *on the normal curve as illustrated below?*

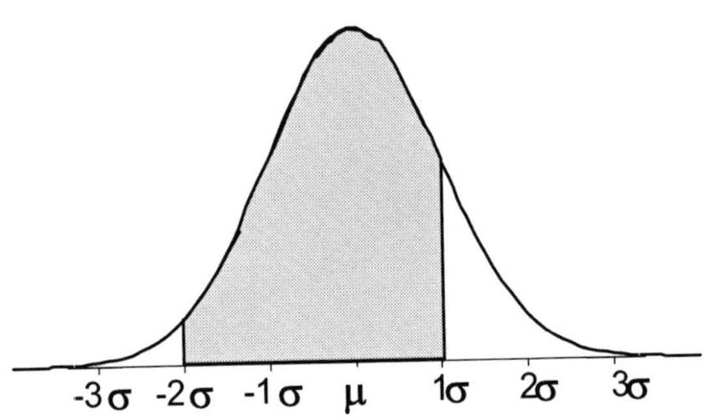

2. *What is the probability associated with the area from* -1σ *to* $+3\sigma$ *on the normal curve as illustrated above?*

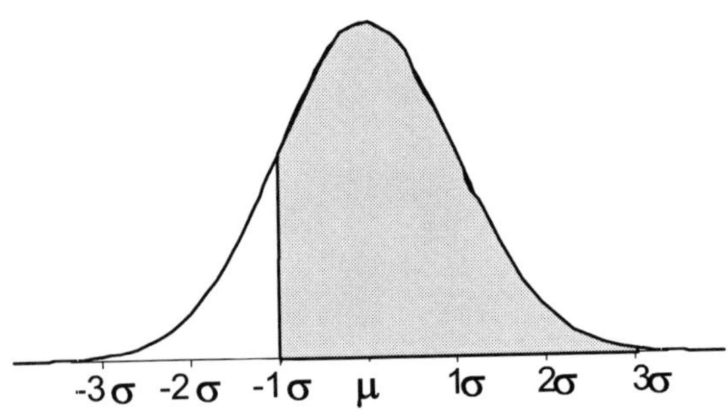

3. *What is the probability associated with the area from $+1\sigma$ to $+2\sigma$ on the normal curve as illustrated below?*

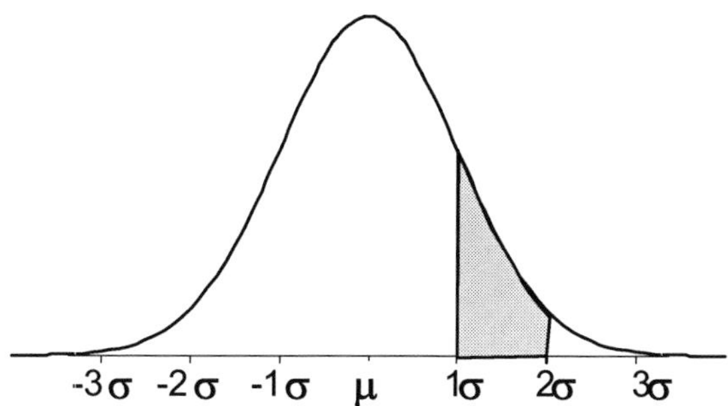

1. *The probabilities associated with the intervals of the normal curve can be determined through simple addition and subtraction (as long as the intervals are multiples of σ). In the case of the interval from -2σ to $+1\sigma$: notice that if we divide the curve in half at the mean, the left portion is just half of the interval -2σ to $+2\sigma$; and the right portion is half the interval from -1σ to $+1\sigma$. The value of the probability is therefore:*

   for the left portion $= \frac{95\%}{2}, = 47.5\%$

   for the right portion $= \frac{68\%}{2}, = 34\%$

   the total probability $= 47.5\% + 34 = 81.5\%$

2. *For the interval -1σ to $+3\sigma$ the calculations follow the same logic as the preceding calculation. Notice that if we divide the curve in half, the left portion is half the interval from -1σ to $+1\sigma$, and the right portion is half the interval from -3σ to $+3\sigma$. The value of the probability is therefore*

   for the left portion $= \frac{68\%}{2}, = 34\%$

for the right portion $= \frac{99.75\%}{2}$, *= 49.875%.*

the total probability = 34% + 49.875% = 83.875% or rounded to 83.88%.

3. *The value of the probability from* $+1\sigma$ *to* $+2\sigma$ *is a little more complex to envision. Instead of addition this will involve some subtraction. It still relies on the symmetry property of the normal curve. Previously we have added a left portion of the curve to a right portion. In this instance only half the curve need to be dealt with. So for a starting point, envision only the right half of the normal curve. The interval we want to calculate is the (right half) probability associated with the interval* μ *to* $+2\sigma$*, minus the probability associated with the interval* μ *to* $+1\sigma$*. The values are:*

   for the first interval $= \frac{95\%}{2}$, *= 47.5%*

   for the second interval $= \frac{68\%}{2}$, *= 34%.*

   the result of the subtraction is therefore: 47.5% - 34% = 13.5%

What about probabilities associated with values like μ to $+1\frac{1}{2}\sigma$? At first thought this would entail adding half the probability difference from $+1\sigma$ to $+2\sigma$ to the value for μ to $+1\sigma$. However, this would give us an incorrect value. For calculation of probabilities associated with points on the normal curve other than $\pm1\sigma$, $\pm2\sigma$, or $\pm3\sigma$ we must turn to the “Z” score and the “Z” table.

"Z" Score or Standard Score

Previously we developed a population of ages that had a mean value of 35 years, and a standard deviation of 6.3. If we were to develop a normal curve for such a population the result would look like:

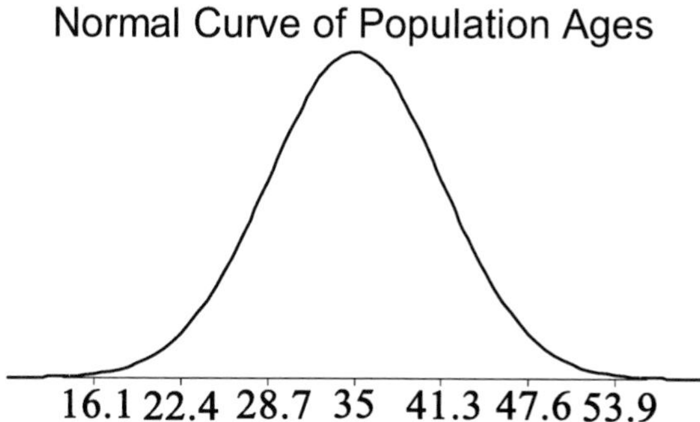

The mean at 35 clearly relates to the middle of the curve. However, to map any other population age to the curve would be difficult because of the decimal number values at the points equivalent to $\pm 1\sigma$, $\pm 2\sigma$ and $\pm 3\sigma$. This curve would not be useful for mapping any other population or distribution because it is uniquely configured to one population.

The "Z" score or the standard score (they are the same thing) is the result of expressing population values as the number of standard deviations above or below the mean (like $+1\frac{1}{2}\sigma$). The "Z" score is a dimensionless number, which means it is a pure number with no units attached (it is not in years, dollars, inches, etc.). Thus the use of "Z" scores allows very different measurements to be compared - kind of like being able to compare apples and oranges, all with reference to the normal curve.

The calculation of the "Z" score can best be shown by example. The previous population that we measured for age had a $\mu = 35$ and $\sigma = 6.3$ and population values that were: 40, 24, 37, 43, 36, and 30. Given these values, what is the *Z* score for the first population member - age 40? To determine the "Z" score we do two things: first; subtract the mean value from 40,

$$40 - 35 = +5;$$

second; divide the result by the standard deviation,

$$\frac{+5}{6.3} = +0.79$$ (equals the Z score for age 40 for this population).

Notice that this score <u>is</u> dimensionless, it is not 0.79 years it is just 0.79. The formula for Z is rather simple:

Z Scores on the Normal Curve

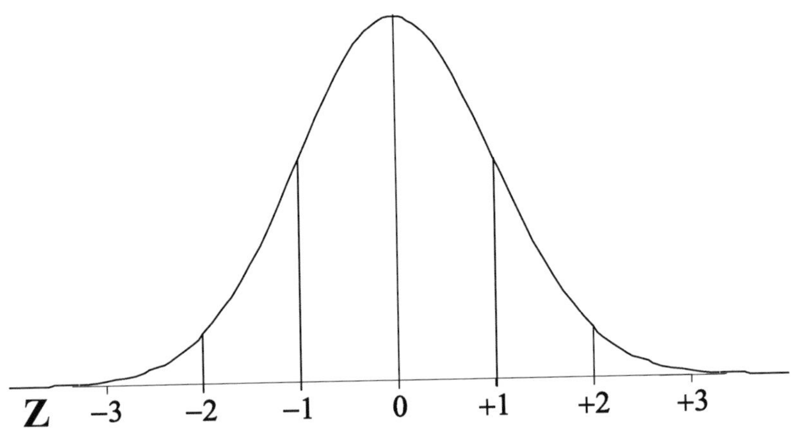

$$Z = \frac{X_i - \mu}{\sigma}$$

Does this formula make sense? Let's work it through. If for a distribution, we had a value that coincided with the distribution's mean, it should "map" to a Z value of zero. Subtracting the mean from this value will result in a 0 value, the result we expect. What about other values? Suppose we had a distribution value equal to one standard deviation above the mean. Subtracting the mean will result in a value of $+1\sigma$, which when divided by σ will result in a value of +1, just the "mapping" we need.

Exercise 5.3
Calculate the Z-scores for the following situations, then for all four Z values rank order them from smallest to largest. (The value furthest left on the Z curve should rank = 1, next to the right rank = 2, etc.)

1. $X_i = 3.9,\ \mu = 4.5,\ \sigma = 0.95$ $Z =$ ____ *Rank*____

2. $X_i = 50.7,\ \mu = 40,\ \sigma = 8.4$ $Z =$ ____ *Rank*____

3. $X_i = 121,\ \mu = 100,\ \sigma = 15$ $Z =$ ____ *Rank*____

4. $X_i = 190,\ \mu = 212,\ \sigma = 13$ $Z =$ ____ *Rank*____

The Z scores you should have calculated should be:

1. $Z = \frac{3.9 - 4.5}{.95},\ = \frac{-0.6}{.95},\ = \text{-}0.63$

The corresponding values (without calculations shown) are:

2. $Z = 1.27$
3. $Z = 1.40$
4. $Z = \text{-}1.69$

If you are unable to obtain the correct answers to numbers 2 through 4, compare your calculations with your learning partner.

To determine the rank order, plot these values on the Z curve as shown following. Rank the values from lowest (furthest left on the curve) to highest (furthest right on the curve). The results are:

1. $Z = \text{-}0.63$ Rank__2__
2. $Z = 1.27$ Rank__3__
3. $Z = 1.40$ Rank__4__
4. $Z = \text{-}1.69$ Rank__1__

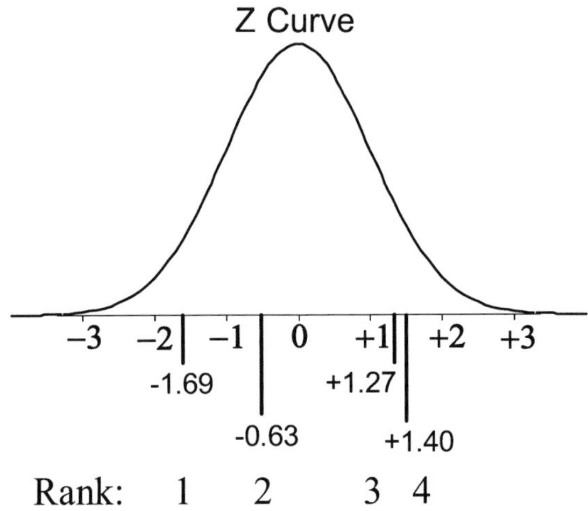

Note that we can tell rank order for numbers from very different population distributions because of the conversion to Z scores. The lowest ranked score (-1.69) was actually derived from the largest value of the raw data: 190. Conversion of data to Z scores allows cross comparison of data from very different population distributions.

Conversion of data into Z scores also allows us to use the properties of the normal distribution to make probability statements. For example with the sample population used previously, what is the probability that a population member of age greater than 40 will be found? We already know that 40 corresponds to a Z value of 0.79, so this question becomes what probability is associated with *Z* values greater than 0.79? To answer this question we must for the first time use the *Z* table. The Z table (Appendix 2, Areas of the Standard Normal Distribution) is an abridged table that lists the probability associated with any value of *Z*. In this table the probability values listed are the probabilities from the mid-point of the curve up to and including the Z value. The column labeled "area" is the probability associated with the given Z value. For our example above we find a table value for Z = 0.79 of 0.2852. (Enter the table at the row for a Z value of 0.7, and track across the row to the column labeled .09 to find the number .2852. This is the probability .2852 or 28.52%

associated with the Z value of 0.79) To convert this to a probability of a Z value less than or equal to 0.79, add the 0.5 for the left half of the normal curve giving a total probability of 0.7852 (.2852 + .5000). What is the probability for a *Z* value greater than 0.79? The answer is

1.0 - 0.7852 = 0.2148.

Thus in the example population, there is approximately a 21.5% probability that an age greater than 40 will be found. Additional examples of calculations of Z scores are contained in Exercise 5.4. For these calculations the use of mirror image values and symmetry of the normal curve is very helpful.

Exercise 5.4

For the following Z scores (calculated in Exercise 5.3), calculate the probability that larger values will be found in their respective population distributions.

1. *–0.63*
2. *+1.27*
3. *+1.40*
4. *–1.69*

To determine some of these values, we must take advantage of the symmetry of the normal curve.

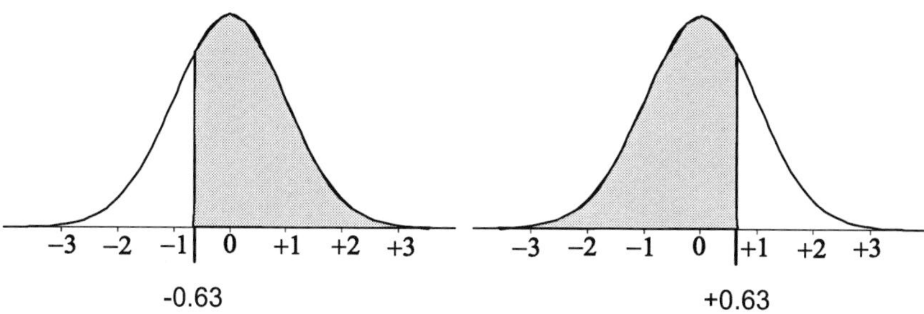

1. For the Z score of –0.63, the question of what proportion of the normal curve is greater than –0.63 can be converted into a mirror image question: what proportion of the curve is less than +0.63? The

illustrations below demonstrate the equality of these statements. So the result from finding the proportion of the normal curve below +0.63 will answer the question. First notice that for the second curve, the left half of the curve represents 50% of the curve. The right half (from 0 to +0.63) can be found directly from the Z table, which is partially reproduced below:

Z	.00	.01	.02	.03	.04 ...
0.6	.2257	.2291	.2324	**.2357**	.2389

Combining the value of the left half of the curve (0.5) and the value of the right segment (.2357) we get a total probability of .7357 or 73.57%. We would state this result as: for this distribution, the probability of a member of the population having a value greater than 3.9 (the original X_i value) is 73.57%.

2. For the Z value +1.27, find the corresponding proportion from the Z table = 0.3980. Since the question is what proportion of this distribution exceeds the (original X_i) value of 50.7, it is equal to the amount of the right half of the curve beyond 0.3980. That value equals 0.5 – 0.3980 = .1020 or 10.20%.

3. For the Z value 1.40, find the corresponding proportion from the Z table = 0.4192. As calculated in problem #2 above the resultant probability for a population member with a value greater than (original X_i) value of 121 = .0808 or 8.08%.

4. For the Z value –1.69, the solution is calculated by the same symmetry properties as # 1 above. The proportion for Z = +1.69 from the Z table is 0.4545. Added to the 0.5 for the left half of the curve (following the same process and logic as # 1 above) we get a total probability of 0.5 + 0.4545 = 0.9545 or 95.45%. Or as stated the probability of a population value greater than (original X_i) value of 190 is 95.45%.

6 Inferential Statistics

Inferential Statistics

Up to this point we have been studying concepts that are largely applicable to the description of populations. The real power of statistics is its ability to predict about a population based on measurements of a sample, a process that we call *inferential statistics*. The preceding work with descriptive statistics prepares us for the process of inference wherein we develop *statistics* for a *sample* and use the information to predict (make an informed guess about) the *parameters* of the *population*. In other words: we are going to use the mean and standard deviation of a sample, along with information from the normal or other distributions, to be able to predict the mean and standard deviation of the population. In order to do this process we have to learn concepts of sampling, calculation of statistics for a sample, and further study the relationships of the normal distribution.

Creating a Sample

The emphasis in describing a population was to develop a clear distinction about the boundaries separating the population from the rest of reality. It is also very important that others be at least conceptually able to recreate the same population based on our (operational) definition. The process of drawing a sample from the population is a little easier. The whole concept of a *sample* is that it represents the population, but it does not have to exactly reproduce it. No two samples are expected to be the same either in individual values of the sample items or in their mean and standard deviation. Instead each sample is considered to be representative of the population from which the sample is drawn. How do we assure that the sample well represents the population? In order for the relationship between the sample and the population to be used for inferences it is important the sample be as *representative* as possible. The best way to assure a representative sample is to utilize a random sample. A *random sample* is a sample developed in such a way that every member/element of the population has the same probability to be selected into the sample as any other member/element.

A random sample is not necessarily easy to obtain. To ensure all elements of the population have an equal chance to be selected, the population must be well known. Because it may often be difficult to have sufficient knowledge about a population, arbitrary samples or samples of convenience are obtained as a best effort towards a random sample. However, just because a sample is arbitrary does not mean it is random. If sufficient information about the population is known, the random sampling process is employed often using a random number table (see Appendix 1) or other random number generator to select members from the population. If it is known that different sub-populations exist in the population (different shifts, distinct age groupings, classes, etc.), a *stratified* sample is drawn. In such a process the sample is broken down proportionally into sub-samples each of which is drawn randomly from within the appropriate sub-population.

The issue of effective sampling is critical to inferential statistics. A poorly developed sample can lead to incorrect or biased inferences about the population. The process of sampling relies on an effective operational definition of the sample - how did you acquire the sample, was it truly representative? In most cases the cost of measuring an entire population is prohibitive. The value of inferential statistics is that a well-developed sample will allow you to develop nearly the same information (knowledge) about the population at far less cost.

Exercise 6.1

Define a population that is familiar to you from your work or school environment. Describe how you would develop a representative sample from this population. What issues would you pay attention to? What concerns would you have? Share your results with your learning partner.

____

____

____

____

____

____

____

____

____

The issues to be considered for almost any sample include:

- *How big a sample would I like to take?*
- *Is it possible to prepare a random sample?*
- *Is stratification necessary? How would stratification need to be performed?*
- *If a random sample cannot be acquired, what is the best sample of convenience or arbitrary sample that can be obtained?*

Sample Distributions

Because a sample is a smaller, representative (hopefully) chunk of reality taken from a population, each sample taken will have its own unique *sample distribution* which will mimic the population distribution, but seldom will exactly reproduce it. The more representative the sample is, the more the sample distribution will resemble in appearance the population distribution. Not only will the sample distribution mimic the population, but the sample mean will also mimic the population mean. Any given sample will have a mean that will likely differ in some amount from the population mean. The difference between a given sample mean and the population mean is termed *sampling error*. It is an error when a difference exists between the sample mean and the true population mean. It is an error, but it is not a mistake, the term "error" in this case means difference from the true value. These errors should be random, both in direction and amount. To the extent the errors are random, they reflect the inevitable fact that sample statistics represent but do not exactly equal population parameters. If the errors are consistent in either direction or amount they are called *biases*, and are reflective of a mistake either in sampling or in the measurement process used for the samples. For example, heights measured with a tape measure that has two inches broken off the end will produce a consistent two-inch bias in the resultant heights. The fact that the general population considers itself above average in almost any measure (whereas from the normal distribution we realize that half the population is below average) would be an example of a bias in a consistent direction. A random sampling process is one of the best ways to avoid biases resulting from sample mistakes.

Sample Statistics

When we generate measures of central tendency (the mean, median or mode) or of dispersion (the range, variance or the standard deviation) for a sample we call the result a sample statistic. This term *sample statistic* is the equivalent for samples as the term population parameter is for populations. The mode, median and mean of a sample are calculated identically to these values for a population. However, the sample mean is abbreviated with a different symbol: $\overline{X}$, pronounced "X-bar":

$$\overline{X} = \frac{\sum X_i}{n}$$

Another different symbol when dealing with sample means is the symbol for the sample size, n, the size of the specific sample being examined.

For each sample that we calculate the sample mean $\overline{X}$, we can expect a (slightly) different value than the $\overline{X}$ for a different sample. A very important fact about the sample mean is that if enough samples are drawn from the population and a mean is calculated for each sample, then the average of each of these different sample means will (in the long run) exactly equal the population mean. This relationship between means of samples and population mean is stated in the language of statistics by saying that the sample mean is an *unbiased estimator* (correct guess) of the population mean. Thus while any individual sample mean will seldom exactly equal the population mean, in the long run with enough samples, the average (mean) of the means exactly equals the population mean.

An equally important sample statistic is the sample standard deviation. The standard deviation of a sample is abbreviated with lower case letter "s". It is calculated slightly differently than the population standard deviation. Were they calculated with the same formula, the sample standard deviation, s, and the population standard deviation, σ, would not have the unbiased estimator relationship. The reason relates to the fact that the standard deviation is more strongly affected by extreme values of the population (large differences contribute more when squared than small differences) than by values nearer the middle of the distribution. The population standard deviation, because it is calculated using all values for the population, totally reflects this disproportionate influence of the extreme values.

The sample standard deviation, on the other hand, is only calculated from the population values present in the (particular) sample. Because the great majority of the population is more likely distributed nearer the center of a population, a sample is more likely to have elements from near the center of the population than from the extremes. (Since a random sample is selected so each population element has an equal chance of being selected, most of the sample will come from nearer the center of the population where most of the population is.) Because the sample will tend to have less extreme values from the population, the sample standard deviation calculated for the sample will consistently tend to underestimate the population standard deviation.

It is possible to compensate for the effect of this underestimation. If we calculate the sample standard deviation, s, by dividing the sum of the squared deviations by "n-1" instead of by n, then s becomes an unbiased estimator of σ. ("n" is the symbol for the sample size as opposed to "N" the symbol for the population size) Why do we use n-1 to divide the sum of the squared deviations instead of n? First, since s (if uncorrected) tends to underestimate σ, we need for our correction to make the (uncorrected) value of s larger. Dividing a number by a smaller number makes for a larger result. Subtracting 1 from n makes the divisor smaller. Thus this correction is increasing the (corrected) value of s to a larger number as required to compensate the underestimation. Second, consider that the smaller the sample size, the more likely that the sample will not have adequate representation of the population extremes. As the sample size increases, the amount of necessary correction will tend to be smaller because more inclusion in the sample from the population extremes is likely. Subtracting 1 from the sample size as a correction factor means that for a small sample the relative correction is larger than for a bigger sample. For instance 5-1 = 4, a 20% correction for this small value of n, 25-1 = 24, a 4% correction to this larger value of n.

How well does the use of n-1 compensate? It compensates exactly! The sample standard deviation, s, calculated using n-1 is an *unbiased estimator* of the population standard deviation, σ. The correct formula for the sample standard deviation is:

$$s = \sqrt{\frac{\sum(\overline{X} - X_i)^2}{n - 1}}$$

Note that this equation is the same as that for the sample standard deviation except for the symbol for the mean ($\overline{X}$ instead of μ) and the use of $n - 1$ as the divisor.

The same corrections apply to calculating values of the variance for samples. The corresponding equation for the sample variance is:

$$\text{variance} = \frac{\sum(\overline{X} - X_i)^2}{n - 1}$$

with the same symbol changes and divisor change as the sample standard deviation above.

Simpler Calculations of Sample Variance and Standard Deviation

The methods for calculating population variance and standard deviation more easily also apply to calculating corresponding sample variance and standard deviation. The divisor value of n – 1 rather than N is also used for these sample calculations:

$$s = \sqrt{\frac{\sum X_i^2 - n\overline{X}^2}{n-1}} \quad \text{and variance} = \frac{\sum X_i^2 - n\overline{X}^2}{n-1}$$

Exercise 6.2
Calculate the measures of central tendency: mode, median and mean; and the measures of dispersion: range, variance and standard deviation for the following samples.

A. Scores of: 4, 5, 3, 7, 6, 6, 5, 4, 5, 7, 4, 6, 6

mode = _____, median = _____, $\overline{X}$ = _____
Range = _____, Variance = _____, s = _____

B. Costs of $24.95, $13.95, $15.95, $12.95

mode = _____, median = _____, $\overline{X}$= _____
Range = _____, Variance = _____, s = _____

Check with your learning partner on the correct calculation of the values of central tendency if your values disagree with these results (or re-calculate your results to obtain these values)

Measures of central tendency:

A. *mode = 6, median = 5,* $\overline{X} = 5.2$
B. *mode = none, median = $14.95,* $\overline{X} = \$16.95$

Measures of dispersion:

A. *range = 7-3 = 4*
B. *range = $24.95-$12.95 = $12.00*

Variance and standard deviation:

A.

i	X_i	$(\bar{X} - X_i)$	$(\bar{X} - X_i)^2$
1	4	+1.2	1.44
2	5	+0.2	0.04
3	3	+2.2	4.84
4	7	-1.8	3.24
5	6	-0.8	0.64
6	6	-0.8	0.64
7	5	+0.2	0.04
8	4	+1.2	1.44
9	5	+0.2	0.04
10	7	-1.8	3.24
11	4	+1.2	1.44
12	6	-0.8	0.64
13	6	-0.8	0.64

B.

X_i	$(\bar{X} - X_i)$	$(\bar{X} - X_i)^2$
24.95	-8.00	64.00
13.95	+3.00	9.00
15.95	+1.00	1.00
12.95	+4.00	16.00

A.

$$\sum \frac{(\bar{X} - X_i)^2}{n-1} = \frac{18.32}{13-1}$$

$$variance = \frac{18.32}{12}$$

$$variance = 1.53$$

$$s = \sqrt{\frac{(\bar{X} - X_i)^2}{n-1}} = \sqrt{1.53} = 1.24,$$

B.

$$= \frac{90}{4-1}$$

$$= \frac{90}{3}$$

$$= 30.0$$

$$= \sqrt{30.0} = \$5.48$$

Degrees of Freedom

The members of a population can have as many differing or even unique values as there are members of the population; up to the population size "N". There are no restrictions on the values for the population, the population parameters μ and σ reflect all population values and are said to have "N" *degrees of freedom*. (Another way to say this is that populations are what they are, there are no restrictions on a population's values.) Consider a sample that is to represent the population mean. In order for the mean for the sample to represent the population mean, there is a restriction on the number of sample values that are free to vary. If the sample mean $\overline{x}$ is to be an unbiased estimator of the population mean μ then all but one of the sample values can vary. The last sample value is not free to vary; it will have to be a specific value, in order that the differences of all the sample values from the sample mean still add up to zero. We call this restriction the "degrees of freedom" of the sample, and for most simple statistics it is equal to one less than the sample size, or "n-1". This value of "n-1", the degrees of freedom of a sample, is the same "n-1" used in the calculation of a sample standard deviation. So measures of dispersion for both populations and samples are really based on degrees of freedom for the respective populations and samples, not just solely the size of the populations or samples. Issues of degrees of freedom beyond "N" and "n-1" are beyond the scope of this text and are addressed in advanced statistics texts.

Sampling Distribution of Means

If we draw repeated samples of the same size from a population and calculate sample statistics such as $\overline{X}$ and s for each sample, we could create a new distribution of the results (for example the means $\overline{X}$'s) for the many samples. Such a distribution of sample means is called a *sampling distribution*. What would the sampling distribution of the sample mean, $\overline{X}$ for those multiple samples look like. Suppose we draw many hundred samples of sample size 10, n=10, from a large population and calculated the mean for each sample. If we formed a distribution of those many sample means, what will the distribution look like? It will be nearly normally distributed, no matter what the distribution of the original population is. *(This is a very key concept that is further illustrated in chapter 8.)* Because the sampling distribution of sample means (for sufficiently large sample size n) will tend to be normal, we can use the probability relationships of the normal distribution to predict population means from sample means, no matter what the original distribution of the population. That is, we can use the power of the normal distribution to make probability statements about the population, even if the population is not normally distributed. This fact that sampling distributions tend to be normal creates the power and capability of inferential statistics (allows us to guess effectively).

Suppose we had developed a large number of samples of size n = 4, and plotted the distribution of sample means in comparison to the original population. (This comparison is made assuming the original population is very large and normally distributed.) How would the two curves compare? Because each sample mean represents an unbiased estimate of the mean (or the middle) for the population, can you imagine that the distribution of sample means is going to be narrower (closer to the center) than the original population distribution? How about if we drew as many additional samples of size n = 25 from the same population? Can you imagine the distribution of means from samples of 25 as likely being narrower than for samples of 4? Why? Because for the

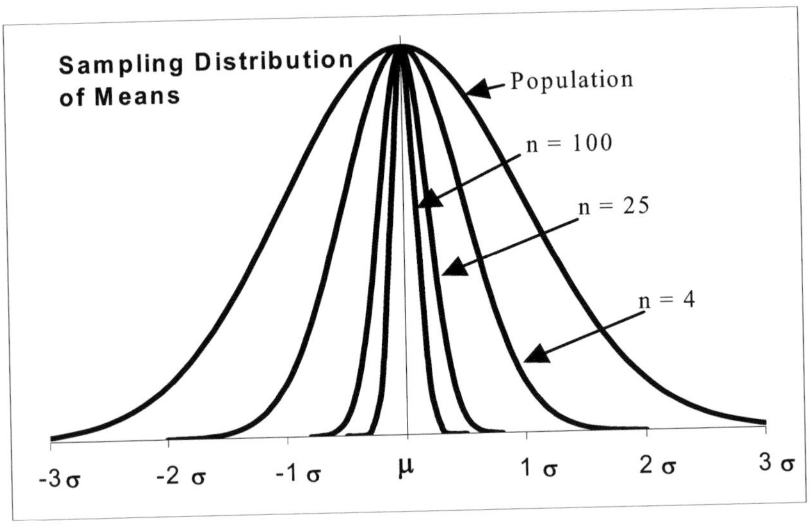

mean of each sample of 25, many more population values are averaged and the result will more closely represent the population center value than the mean of each sample of 4. Examples of distributions of sample means for sample sizes of 4, 25, and 100 are shown above. It is apparent that the larger the sample size, the more tightly the distribution of sample means clusters around the population mean. For each of the respective sample distributions we could calculate a standard deviation (called the standard deviation of the mean and abbreviated $s_{\overline{X}}$). Given that the distribution of means for the various sample sizes (n = 4, n = 25, n = 100) is different; the related standard deviations of these means would also be different. How does the standard deviation of the sample means for the various sample sizes compare with the population standard deviation? As you can see in the preceding illustration, the larger the sample size, n, the narrower the distribution and therefore the smaller the standard deviation of the sample means. As a matter of fact there is a mathematical relationship between the population standard deviation and the standard deviation of means for various sample sizes drawn from that population.

Standard Deviation of Means

Before presenting that mathematical relationship, it is worthwhile to review the new abbreviation for the term "the standard deviations of the means of samples". The abbreviation is a combination of the symbols for a sample standard deviation, s, and the mean, $\overline{X}$. The symbol that is used universally is the combination of the two with the $\overline{X}$ subset to the right of the s. Thus the symbol is, $s_{\overline{X}}$, pronounced "s-sub-X-bar". How do these standard deviations of the means relate to the standard deviation of the population? As demonstrated with the illustration, the larger the sample size, n, the smaller the standard deviation of the mean. The relationship is that the *standard deviation of the mean* is equal to the standard deviation of the population divided by the square root of the sample size, n:

$$s_{\overline{X}} = \frac{\sigma}{\sqrt{n}} \quad \text{or} \quad s_{\overline{X}} = \frac{s}{\sqrt{n}}$$

The second equation is true because the sample standard deviation is an unbiased estimator of the population standard deviation. If the standard deviation has not been measured for the entire population, then you use the best unbiased estimator available, in this case "s". (The standard deviation of the mean, $s_{\overline{X}}$, is also called the *standard error of the mean.*)

What is the significance or value of this relationship, or SO WHAT? How does the relationship of the standard deviation of the mean and the square root of the sample size have practical application for us? One application is when we are trying to predict the mean for a population by the measured mean of a sample. If you look at the preceding curves for the sample size n =100, you see that even mean values on the extreme edge of the curve are quite close to the true population mean. On the other hand, for a sample size n = 4 the more extreme values of the sample means can depart quite a bit from the true population mean. Because we will be using the data from samples to predict the population mean, we can improve the "closeness" (or accuracy) of our prediction just by increasing the sample size. If this is the case, why don't we just use very large samples to improve our ability to predict μ?

The creation of samples and measurement of their characteristics usually costs us something: money, time, or other resources. Our ability to predict increases as the square root of the sample size, but these costs usually increase directly with the sample size. For any given sample size we can increase our ability to predict the population mean by a factor of 2 by increasing the sample size (and presumably our costs) by a factor of 4. Thus even in the most important inferential measurements, you seldom see extremely large samples employed. Even the important TV household rating values are prepared from data collected from approximately 1600 U.S. households with the data applied to represent the viewing habits of many hundred million Americans.

For example if we were doing a measurement that cost $10 per individual member of the sample, our cost and ability to predict the population value would be related as follows:

n (sample size)	Cost	Improvement Factor*
100	$1,000	10X
500	$5,000	22X
3,000	$30,000	55X
10,000	$100,000	100X

*Improvement over predictive ability of a single sample, n = 1

There is another application of this relationship. It is used when we calculate a mean for a sample, and want to use that mean not just to represent (guess) the population mean, but rather to develop a range of values that have a high probability of including the true population value. For that calculation we will use the standard deviation of the mean and the properties of the normal curve. The range of values created is called a confidence interval. Confidence intervals will be discussed in later chapters.

Calculating Z Scores with Sample Values

When we previously studied Z scores for values, we used population parameters μ and σ for the calculation. When we are working with sample data, we usually do not know these values and instead substitute the corresponding sample values $\overline{X}$ and s as unbiased estimators of the population values. The corresponding equation for the Z value is:

$$Z = \frac{X_i - \overline{X}}{s}$$

To summarize key aspects of inferential statistics:

- Inferential statistics apply to measures developed from a sample.
- In general, Roman letters are used for symbols for sample values: $\overline{X}$, s, n, p.
- The collective term for measures and values that apply to samples is *statistics*. An example of the usage of this term is: Values of $\overline{X}$, s, n and p are statistics for a given sample.

7 Correlation/Regression and Statistical Tests

Correlation and Regression:

Often when we are studying the data for populations or samples we are interested in comparing two items of information for each sample or population member. Examples of such comparisons include: age and net worth, length of marriage and happiness rating, automobile weight and gas mileage, etc. Such a comparison is begun visually using an x-y plot or what is often called a scatter diagram. This is a method of putting the information onto a two-dimensional representation where the two items of information are scaled respectively onto the "x" or horizontal axis and the "y" or vertical axis. Convention suggests that the independent variable (or controlling variable) is plotted along the x-axis while the dependent (or controlled) variable is plotted along the y-axis. Such a plot presents a visual impression of the relatedness of the two items. Often we would like to develop a number or set of numbers that will measure the inter-relatedness. Regression and correlation are two analyses producing numbers that are used for the evaluation of inter-relatedness.

Regression is the process of finding a straight line which best fits the scatter diagram data. The "best fit" means for the straight line developed by the regression equation, the total of the differences from data points to the line is minimized (technically the differences are squared for the same reason we squared differences in calculating standard deviations). For most sets of data only one such line exists and can readily be calculated. Such a line is usually expressed as a formula with y values for the line expressed as a function of x values. The usual representation of the formula is $y = a + b(x)$, where the two numbers represented by the "a" and "b" are determined as a result of the regression calculation. The three examples of correlation that follow also show the respective line of "regression" for the correlated data.

Regression gives a straight line which best fits the data, however, regression gives no information about *how well* the line fits the data. The *correlation coefficient* is a number (symbolized by "r") that indicates how well the regression line fits the data, or since it is a calculation separate from regression calculation, how strongly the data are linearly interrelated. The correlation coefficient is a number varying from -1 to 0 to +1. The closer the correlation value approaches either +1 or -1, the stronger the interrelationship between the x and y variables. The closer the correlation value is to 0, the less the interrelatedness (a correlation value equal to zero indicates no linear interrelatedness whatsoever). The sign of the correlation coefficient (plus or minus) indicates the direction of the interrelationship: a plus indicates that as x increases so does y, a minus indicates that as x increases the values of y decrease. (For example, gas mileage of automobiles would likely decrease as weight increases resulting in a negative correlation coefficient, as illustrated on the next page.)

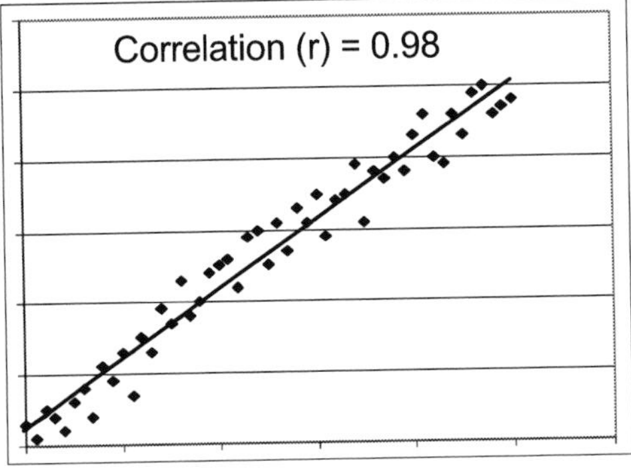

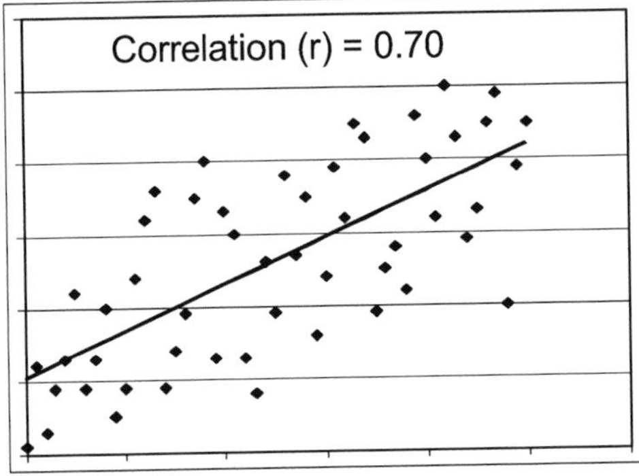

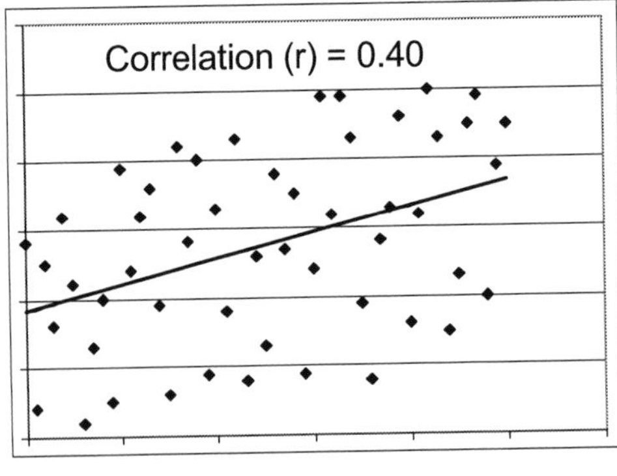

How do you interpret the correlation coefficient? As shown in the examples at left, for values near zero, little or no relationship is indicated. For values from ±0.3 to ±0.5, a tenuous relationship is indicated; from ±0.5 to ±0.7, much more definite relationship is indicated; and for values from ±0.7 on up to ±1 a very strong relationship is indicated. Suppose that a set of data was found to have a

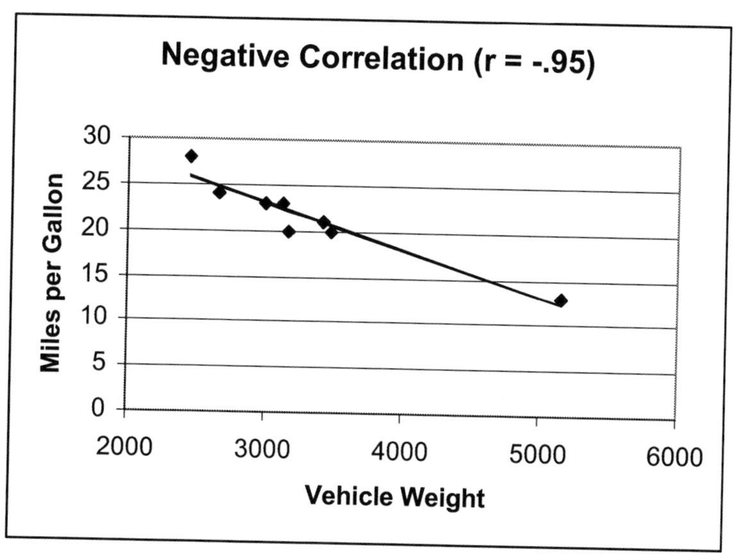

correlation of positive 0.98 as illustrated, what conclusions could one draw? The only conclusion that may be legitimately drawn is that the two variables show very strong interrelatedness. A high correlation coefficient does not indicate causality. One of the most common misuses of statistics is considering a high correlation coefficient as a proof of cause and effect. While a cause and effect relationship may exist, a high correlation coefficient does not "prove" anything. The apparent interrelatedness of the two variables may in fact be due to a common causality from a third variable. In those instances there is a good reason to assume causality does exist, the square of the correlation coefficient is indicative of the percentage of interrelatedness. In the example of the vehicle weight vs. miles-per-gallon above, the $r = -0.95$ yields an $r^2 = 0.90$. This indicates that 90% of the variation in miles-per-gallon among the models can be attributed to the variation in weight. This is a very strong interrelatedness. The remainder of the variation (10%) of miles-per-gallon can be attributed to other factors: type of transmission, engine technical design, streamlining of the vehicle shape, etc.

Exercise 7.1

Answer the following questions regarding correlation and regression to see if you understand these issues:

1. A newspaper interviews a professor of statistics about a proposed system for rating the teaching ability of faculty members. The professor says "The evidence indicates that the correlation between a faculty member's research productivity and teaching ratings is close to zero." The paper reports this as "The professor said that good researchers tend to be poor teachers, and vice-versa" Why is the paper's reporter wrong? Write a short explanation of the professor's true meaning and share it with your learning partner.

2. Each of the following statements contains a blunder. In each case explain what is wrong.

- *"There is a high correlation between the occupation of American workers and their income"*
- *"We found a high correlation (r = 1.09) between direct reports' ratings of their boss's performance and ratings made by the boss's peers."*
- *"The correlation between planting rate and yield of wheat was found to be r = 0.23 bushel"*

1. The reporter apparently confused a near-zero correlation coefficient with a negative correlation coefficient. The statement should emphasize that a faculty member's research performance and teaching performance are not reflecting the same type of skills and therefore are not interrelated, as indicated by the zero correlation coefficient. A faculty member could be both an excellent teacher and researcher, or excellent or poor at either independent of the other.

2. The errors in these statements are:

- *Worker's occupation is nominal data; it cannot normally be used to create correlation values because interval or ratio data is required to calculate correlation.*

- *Correlation values range from 0 to 1 (either plus or minus), a value of 1.09 is not possible.*

- *Correlation (like the "Z" score) is a dimensionless number, that is a "pure" number without any units attached to it.*

Concepts of Testing

Many people have the idea that statistical tests are used to prove or disprove facts or theories. This idea is wrong! Statistical tests are used as a basis for accepting or rejecting predefined concepts (called hypotheses) with the understanding that such tests are *always* subject to error. The rest of this chapter will be spent describing the types of guesses or hypotheses that are tested, and how the element of error is dealt with.

Basically statistical tests are always applied to an either/or type of situation. For instance we could say that either our class mean age is equal to the mean age of all statistic students, or it is different. Thus whenever a statistical test is being made, these two hypothesis are being evaluated: 1) the data do not (materially) differ from each other, or 2) the data do differ and the difference is real. These options have been given names and definitions in the language of statistics. The *Null Hypothesis* is the hypothesis (a fancy word for a guess) that no real difference exists between whatever (two samples, a sample and a population, etc.).When the null hypothesis is being tested, the assumption is that any apparent differences are due to random errors in sampling. The null hypothesis is symbolized by a capital H with a zero subset to its right (H_0), H standing for hypothesis and the subset zero meaning no difference really exists. ("Null" is the German word for the number zero.) The *Alternate Hypothesis* also called the *Research Hypothesis* holds that the differences that exist between the elements being tested are real. The research hypothesis is symbolized by a capital H with a subset number one to the right (H_1). As above the H is shorthand for "hypothesis" and the subset "one" reflects the alternate character of this hypothesis to H_0. Those familiar with computer theory recognize the use of 0 and 1 represent a "binary" situation – meaning that only two alternatives are being considered. Statistical hypotheses are binary in that as the result of the test one of the hypotheses is accepted, and the other is rejected.

How does one go about making a statistical test? First and foremost is to determine the null hypothesis that will be tested and to understand that only the null hypothesis is ever tested by a statistical test. The research hypothesis is never tested itself; instead it is accepted or rejected based on the results of the test of the null hypothesis. Perhaps an example would help clarify. Suppose that I were comparing the dropout rate for one of my statistics classes with the dropout rate for a conventional statistics class at another institution. In my heart I want the results to suggest that a different (read better!) dropout rate exists for this class. If I had dropout data for each class I might want to compare them using an appropriate test. Before performing the test I would need to formulate the null hypothesis that would be something like this, H_0: no significant difference exists between the dropout rate for the other class and mine. Associated with this H_0 would be the following research hypothesis, H_1: a significant difference does exist between the dropout rate for this and the other class. Again I wish to emphasize that H_0 alone is subject to test and we never test H_1. We will discuss the outcomes of tests and associated errors that can be made in the following sections

Exercise 7.2

Create a statement of the Null hypothesis and the Alternate Hypothesis for each of the following situations. Share with your learning partner.

1. *A counselor who specializes in treating depression has developed a new treatment. He arranges with a colleague to compare his results (as measured by a standard scale of depression) with the colleague's results using a classical treatment. What are the H_0 and H_1 statements?*

2. *A manufacturer puts a new production process in place in parallel to the prior process. Before discontinuing the prior process, a series of production runs are made using both, recording material consumption and labor hours required by both. For both types of data, create the H_0 and H_1 statements. What result is the manufacturer probably most interested in, H_0 or H_1?*

1. *Examples of the statements are:*

   H_0: There is no significant difference between the outcomes obtained using the new treatment versus the classic treatment.

   H_1: There is a difference between the outcomes for the new and the classic treatment.

Which outcome does the counselor desire? It depends. Accepting H_0 means that the new method is as least as good as the classical treatment. On the other hand, rejecting H_0 and accepting H_1 has two possible implications (two-sided test outcome), the first is that the new treatment is less effective than the classic treatment, the second is that the new treatment is more effective than the classic treatment. Use of one-sided tests would allow resolution of these different implications.

2. *For material consumption:*

H_0: There is no difference in the amount of material consumed in the two processes.

H_1: There is a difference in the amount of material consumed in the two processes.

For labor hours utilized:

H_0: There is no difference in the labor hours utilized between the two processes.

H_1: There is a difference in the labor hours utilized in the two processes.

Again the manufacturer is interested in specific implications of the outcome of these tests. The best situation would be results (H_1) for both tests that imply the new process is more conservative of both material and labor hours. It is conceivable that only material or labor hours would be reduced and still be a positive outcome. It is also possible that neither is reduced because the new process is focused on another outcome: much finer quality of product, lower energy utilization, etc.

Levels of Significance and Confidence Intervals

Before discussing the basics of specific statistical tests it is important to understand the probabilities associated with testing. As discussed before, we are always testing the null hypothesis, usually out of a desire to be able to accept a research hypothesis. These tests are not based on absolute certainty, they relate to probability values. The basis for rejecting or accepting the null hypothesis is to predetermine a value of probability that becomes a cutoff point for accepting or rejecting the null hypothesis. We call this cut-off point the *level of significance* (it is symbolized by the Greek letter α). It corresponds to the region at the extremities (or extremity) of the normal distribution curve where there is a very low probability (equal to α) of H_0 results being found. The null hypothesis states that no real difference exists between the results being tested; and that any apparent differences can be accounted for by the probabilities of the sampling process. When we determine the level of significance to apply to a test we are setting the corresponding cutoff point, as illustrated below. The level of significance is inversely related

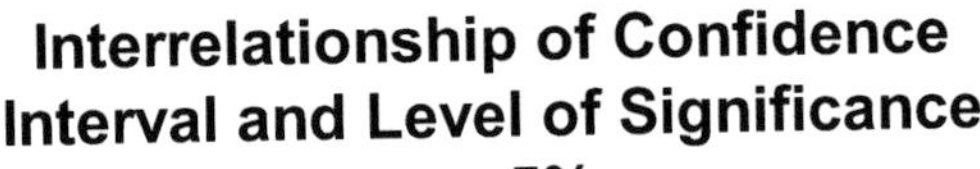

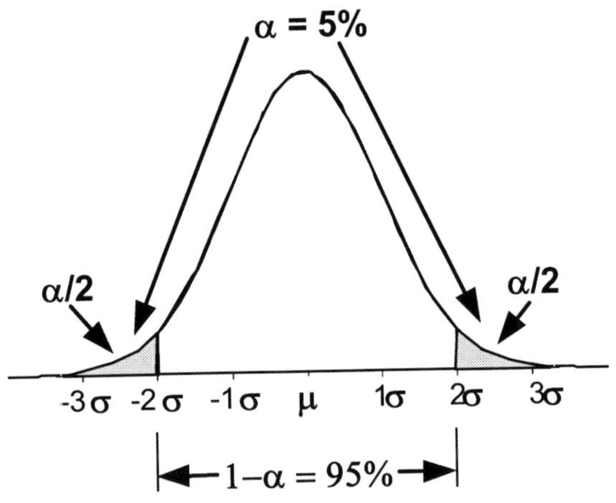

to the confidence interval as shown in the illustration. *Confidence intervals* are created to relate the results for the mean of a sample to a prediction of the corresponding population mean with a known probability equal to $1-\alpha$. Thus the relationship is that the confidence interval (as a probability) equals one minus the level of significance.

The level of significance is a probability expressed as either a decimal fraction or as a percentage, .05 or 5% for example. It establishes a cutoff point where in the test of the null hypothesis, a value of differences beyond the cutoff allows rejecting H_0. This cutoff point typically represents a rather small probability value, because if there is a large probability of a given difference existing between equivalent samples we will be wise to accept the null hypothesis. The most commonly used levels of significance are 10%, 5%, and 1%. Remember from the nature of the normal distribution that these probability values can be associated with extremities of the normal curve and the distance of values from the mean expressed as standard deviations. Another way of saying the same thing is that the "Z" score of the normal distribution can be associated with the probabilities at the extremities or "tails" of the normal distribution. What *Z* values correspond to levels of significance of 10%, 5%, and 1%? From the Z table you can determine the values: $\pm 1.65\sigma$, $\pm 1.96\sigma$, and $\pm 2.57\sigma$ respectively. Thus if we had pre-determined our level of significance to be 5%, (as illustrated above) and we found that the difference between our values was greater than $\pm 1.96\sigma$, then we would reject the null hypothesis and say that the difference was significant.

Note that the above "Z" values associated with the three levels of significance are values where the difference could be at either end of the normal curve, thus the $\pm$ Z value (creating what is called a *two-sided or two-tailed test*). These Z values are calculated by ascribing half the probability related to the level of significance to one end of the normal curve and calculating the corresponding Z value. In the examples above, 10% was evaluated as 5% at one end of the curve, 5% as 2.5%, and 1% as 0.5 %. If we have reasons to expect any differences to occur at only one end of the normal curve, we would perform a *one-sided test* by calculating the Z values for the full level of significance. Thus for the above three levels of significance as a one-sided test on the right side of the Z curve, the corresponding Z values would be: (from the Z table) +1.29, +1.65, and +2.33.

What are confidence intervals about? When we discussed the effect of sample size on the variation of distributions of sample means, we created the term: standard deviation of the mean:

$$s_{\bar{X}} = \frac{\sigma}{\sqrt{n}} \quad \text{or} \quad s_{\bar{X}} = \frac{s}{\sqrt{n}}$$

Suppose that we have a sample of size "n" producing a mean value "$\bar{X}$", with a measured standard deviation for the sample "s". Using this data we can create a confidence interval for the sample results that predict with a known probability of success the population value "μ". Suppose that the sample size is n = 25, the sample mean for this sample of 25 is $\bar{X}$ = 67.5, the sample standard deviation s = 10.1, and the (previously known) population standard deviation is 9.3. What would be a 95% confidence interval resulting from these data. The standard deviation of the mean in this instance would be

$$s_{\bar{X}} = \frac{\sigma}{\sqrt{n}}.$$

When we substitute actual values in the equation,

$$s_{\bar{X}} = \frac{9.3}{\sqrt{25}}, = \frac{9.3}{5} = 1.86.$$

Just as with the level of significance, the Z value for a probability of 95% (α = 5%) is 1.96. Thus the confidence interval for the sample mean is $\bar{X}$ ± 1.96 $s_{\bar{X}}$. The result is

67.5 ± 1.96 (1.86) = 67.5 ± 3.65,

or expressed as a range 67.5 – 3.65 to 67.5 + 3.65

yielding the 95% confidence interval 63.85 to 71.15. How do we interpret this result? It can be expressed as: there is a 95% probability that the true population value μ is between 63.85 and 71.15.

What if the population standard deviation is unknown (a more likely possibility)? Then the sample standard deviation "s" is substituted as an unbiased estimator of σ. The appropriate equation is now

$$s_{\bar{X}} = \frac{s}{\sqrt{n}}$$

$$s_{\overline{X}} = \frac{10.1}{\sqrt{25}}, = \frac{10.1}{5} = 2.02$$

Applying the same Z values as above, the result is

$67.5 \pm 1.96\ (2.02) = 67.5 \pm 3.96$,

or expressed as a range 67.5 – 3.96 to 67.5 + 3.96 yielding the 95% confidence interval 63.54 to 71.46[3].

Exercise 7.3

For the following data, calculate the corresponding confidence intervals.

Sample values: $\overline{X} = 19.4$*,* $n = 17$*,* $s = 2.3$*, (population value)* $\sigma = 2.12$*. Calculate the 90%, 95% and 99% confidence intervals.*

The first activity is to calculate $s_{\overline{X}} = \frac{\sigma}{\sqrt{n}}$*. Since we know* σ *(probably from other studies), we can use it for the calculation of the standard deviation of the mean.*

$$s_{\overline{X}} = \frac{2.12}{\sqrt{17}} = \frac{2.12}{4.12} = 0.51.$$

Then we will perform the confidence interval calculations for each value of $1 - \alpha$*:*

$1 - \alpha$	*Z value*	$\overline{X} \pm Z\,(s_{\overline{X}})$	$\overline{X} - Z\,(s_{\overline{X}})$ *to* $\overline{X} + Z\,(s_{\overline{X}})$
90%	*1.64*	*19.4 ± 0.84*	*18.56 to 20.24*
95%	*1.96*	*19.4 ± 1.00*	*18.4 to 20.4*
99%	*2.57*	*19.4 ± 1.31*	*18.09 to 20.71*

Looking at these results, several issues should be apparent. The first is that the greater the level of confidence (1 – α*), the larger the associated confidence interval. If we want to be really sure about the population*

[3] The use of Z = 1.96 for a 95% confidence interval when using a sample standard deviation is only correct for very large sample sizes. For smaller sample sizes, the value of "t" for 95% is used.

mean (very low α), then the range of values that "contain" the true population value will be wider. What if we wanted greater certainty without increasing the spread of values? The only way to accomplish this is to increase the sample size. A larger sample size will produce a smaller value for the standard deviation of the mean, thus a narrower confidence interval will be produced for the same value of α.

Errors Associated with Tests

One way to think about the level of significance is to consider it to be the probability of a certain kind of error that we might make when we test the null hypothesis. Suppose that we gave a statistics student a standardized statistics test at the beginning of the quarter, and another test at the end of the quarter. Suppose the mean score for the first test was 300 and for the second test was 550, and we further know that for this test the σ is 100. How could we compare the two results and develop a test of the significance of the differences? First, we develop a statement of the null hypothesis, H_0: no significant difference exists between the first and second set of test results. Second, we determine the level of significance we will use to apply the test; in this case we select 5%. Third, we convert the differences between the scores to a "Z" value (ignoring several subtleties which will be picked up later) and we get Z =

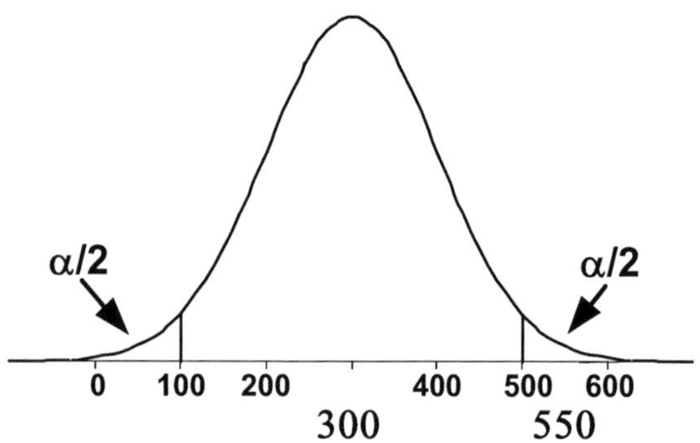

2.50. With our predetermined level of significance of 5% we would reject the null hypothesis since the calculated Z value exceeds the cutoff Z value of 1.96.

But suppose that the difference is really just due to the probabilities of sampling, and we drew the one out of twenty sample? Based on the test outcome, we have just rejected a null hypothesis that we should have accepted. We have just made an error in the results for this test. Can we calculate the probability that we will make such an error? If our level of significance is 5%, then in 5% of the tests performed, such an error will

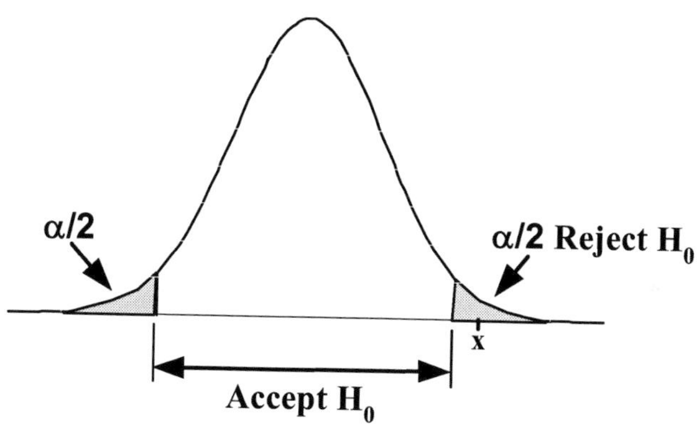

occur. Thus the level of significance, which is symbolized as "α" is also the probability that we will make an erroneous conclusion to our test and reject a null hypothesis that we should have accepted. This is called a "Type I Error" or an "Error of the First Kind". In the example normal distribution preceding, the region associated with α is labeled (for a two-sided test) and a Type I Error is illustrated. Note that the value 550 is in the reject region even though it is legitimately part of the distribution. This is illustrated generically in the normal curve below. Just as there are always two hypotheses that are being evaluated in each statistical test, (null and alternate/research) there are two kinds of errors that can be made as the result of testing. The error of the first kind we have already discussed. It has the probability α associated with it and results in rejecting an H_0 that we should have accepted. The "Error of the Second

Kind", also known as a "Type II Error", occurs when you have accepted a null hypothesis that you should have rejected. This will happen when two distinct distributions exist (the difference is "real") close enough together that a piece of data or element of the second distribution overlaps into the acceptance region of the first distribution. The probability of the Type II error is symbolized by "β", the Greek letter "beta". The value of probability associated with β is very difficult to evaluate because as illustrated it depends on the value of α selected and the interrelationship of the two distributions.

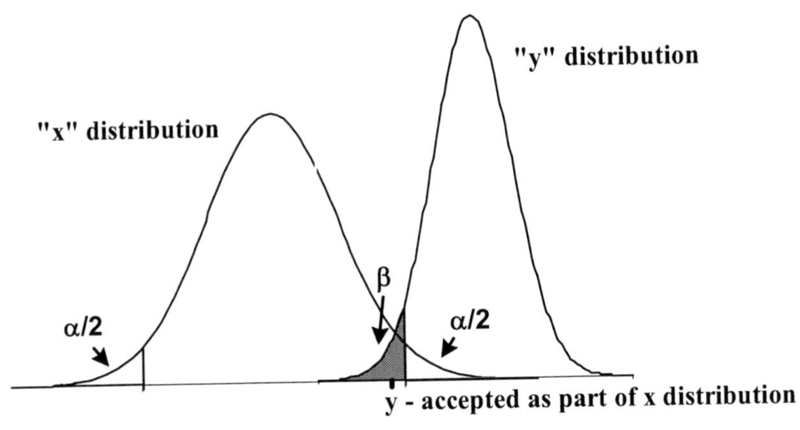

The inter-relationship of α and β shown above explains why when we are going to make a test that we do not just set α to a very low value, say .01%. This would certainly decrease our likelihood of making an error of the first kind (the probability of which is precisely equal to α), however, it would increase (by an unpredictable amount) the likelihood of making an error of the second kind. As can be seen in the illustration above, the same cut-off value defines both α and β. As the cut-off line slides to the right (decreasing α) it is also increasing β. Because the "y" distribution is seldom known, the relationship of β to α is unknown. What is known is that making α much smaller (decreasing the likelihood of an error of the first kind) tends to increase β and the likelihood of an error of the second kind. One of the influences for selecting the level of significance (value of α) is to understand the consequences of making each type of error. Which would have the more disastrous consequences, an error of the first kind or an error of the second kind? If an error of the second kind is more disastrous then we would select a relatively large value of α, say 10%. If an error of the first kind was more disastrous, then we would select a corresponding smaller α, say 2%, 1% or even less. In all cases it is important to remember that there is no certainty whatever the test results, there is always a probability associated with both kinds of error, no matter how we "load the dice" in selecting our level of significance.

Statistical Tests

When we seek to infer from a sample to a population or to show the equivalency of two samples we use statistical tests of the null hypothesis. Before explaining the basis of specific tests, we should understand that there are two general categories of tests called parametric and non-parametric tests. The difference between the two categories is that the *non-parametric tests* are based solely on probability relationships (like a coin toss or dice roll) without any assumptions of underlying population distributions. On the other hand *parametric tests* are based on assumptions about the shape of the underlying population distribution as a normal distribution. Non-parametric tests are therefore more universal in nature because they will apply to all situations. They are less powerful in their application because they are less able to distinguish differences (less sensitive) in testing the null hypothesis. Parametric tests do rely on the properties of a given population distribution (such as the normal distribution) for their power and are correspondingly more sensitive or powerful in testing the null hypothesis. There is a corresponding risk to parametric tests, however, if the data being tested are not from the population distribution necessary to the test, then the test may give false results. Thus before performing the parametric test, it may be necessary to first test the data itself to assure that it is from the assumed population distribution. In fact there are non-parametric tests available to determine whether parametric tests may be used with given data.

The performance of tests always involves a comparison of results for one sample to another sample or from a sample to a population. The comparison, calculated either as a difference measure or as a ratio of results is accomplished using an underlying probability relationship that is the basis of the test. (*See further elaboration of the basis for tests in Chapter 9.*) Thus even though the actual performance of each test is unique, the structure for all tests is similar. A difference or ratio is calculated and compared to a critical value from a table. The value from the table usually reflects both the level of significance α, and the degrees of freedom applicable to the test. Depending on the test structure, if the test calculated value exceeds or in some cases is less than the critical table value, the null hypothesis is rejected.

The sequence of issues to be addressed for performing a test is as follows. First, understand the type of data (level of scaling) you will be testing. Second, formulate the null hypothesis statement. Third, determine the level of significance for the test and whether it will be a one-sided or a two-sided test. Fourth, from the type of data (and for some tests data amount and/or structure) and whatever knowledge you have about the underlying data/population distribution select the test to be performed. Fifth, perform the test. Sixth, based on the results accept or reject the null hypothesis.

Six tests which are commonly used in business and the behavioral sciences will be briefly discussed: Chi-square, sign test, Wilcoxon matched-pairs signed-rank test, t test, F test, and analysis of variance. Details of performing the tests may be found in chapter X and any standard statistical text, thus rather than present formulas here, we will instead seek to understand the applicability of the tests.

Chi Square Test

This is a non-parametric test that is frequently used in the behavioral sciences. It does not require any assumptions about population distributions, and is applicable to data from all four levels of scaling. It reduces all data to nominal counts or proportions of categories of data. The test compares the actual results from a measurement with predicted or theoretical results. It is important in using this test that any predictions that are going to be tested be made before the data is collected. Examples of situations for which this test would be applicable include: comparing marketing survey results with the actual sales results over a period of time; comparing psychiatric test scores for individual patients before and after treatment. This test can also be used to examine homogeneity of data. Effectively the "prediction" is that data are homogeneously distributed across categories, and the test compares the actual distribution to this prediction.

Sign Test

This non-parametric test applies to data from all four levels of scaling. It is applicable only to *paired* data, like before-and-after evaluations, or other pairing schemes. The test is based on the direction (sign) of change of data in these paired environments and treats the direction of the numbers' change basically as nominal data. The sign (+ or -) of the difference in whatever is being evaluated is calculated for each pair. The proportion of pluses and minuses are then compared to what simple probability (essentially like a coin toss where a "head" is a plus change and a "tail" is a minus change) would predict for the H_0 and the level of significance that was determined before the test was performed.

Wilcoxon Matched-Pairs Signed-Rank Test

This is a non-parametric, more powerful version of the above sign test. It is more powerful because it takes into account more of the information then just the sign or direction of change. For the same matched pairs, not only are the signs of the differences used, but also the magnitude of the differences. These magnitudes are ranked from least to most (ignoring the signs), and then the ranks are added together for the positive differences and the negative differences. The summed ranks are then compared to tables, indicating what various probabilities would show as random results for H_0. While this is a non-parametric test, it is more powerful (discriminating) than the sign test because it uses more of the information in the data being tested.

"t" Test

This test is used to compare means between two samples (or more exactly the difference between means) as a test of the null hypothesis. It is a parametric test, assuming that the underlying data is normally distributed. Built into this test is the effect of sample size on distribution of samples. It is similar to the Z statistic except the difference between the means, expressed as number of standard deviations, is compared to a given "t" value for a preset α value and the sample size. If the difference of the means exceeds the "t" value, the null hypothesis is rejected. This test is only applicable to interval and ratio data. Examples of data which could be tested include: height or weight of test samples (people or animals) before and after a nutritionally enhanced diet; average temperatures of the 1980's compared to the average temperatures of the 1800's (are we moving into the ice age?). As the sample size increases the differences in the cutoff values of the t test and the Z statistic diminish. For sample sizes greater than n = 30, the two are virtually indistinguishable.

F Test

This test is used to compare variance (the square of the standard deviation) for different samples or populations. It is a parametric test assuming a normal distribution of the data. It is a relatively easy test to perform as the ratio of the two variances is compared to a table value for a given level of significance and sample size for each variance. This tests the null hypothesis that the two variances are not significantly different. It is often used when additional statistical calculations will require the treatment of the two variances or standard deviations as equal.

Analysis of Variance

One drawback to the t test is that it only operates on the means for two results. What do you do when you want to compare the means for three or more sets of results? The analysis of variance is used for such a comparison. It operates not on the means but rather on a comparison of: the dispersion (variance) within the separate samples against the dispersion among the total samples. Again it is a parametric test that assumes a normal distribution of the sample data. The actual test employs the F test to compare the variance calculated among the samples with the variance calculated between the samples. The calculations are relatively complex and can be found in the chapter 9.

Summary of Statistical Tests

Test Name	*Level of Scaling for Test Metric*	*Test Type*	*Test Statistic Evaluated*
Chi Square	Nominal	Non-parametric	Expected or theoretical proportions (or counts) of nominal data compared to actual
Sign Test	Nominal	Non-parametric	Positive and negative values of changes in paired or correlated data.
Wilcoxon Matched-Pairs Signed-Rank Test	Ordinal	Non-parametric	Paired data, rank ordered by amount and sign (+ or-) of the changes.
t Test	Interval/ Ratio	Parametric	Difference between means for 2 samples and or populations
F Test	Interval/ Ratio	Parametric	Variances of two or more sets of data.
Analysis of Variance	Interval/ Ratio	Parametric	Differences between means for 3 or more samples and or populations

Exercise 7.4
Complete the following and discuss with your learning partner:

- *Describe a situation where you might compare data from two different samples from an aspect of your work or study environment.*
- *Develop an example statement of a null hypothesis and the corresponding research hypothesis.*
- *Describe the consequences of making a Type I Error and a Type II Error for the null and research hypotheses developed above.*
- *Determine the level of significance you would select for the test based on the above error consequences. From the nature of the data, determine which test you would probably use for the above situation.*

Given the open-ended nature of these questions, it is impossible to prepare a set of answers in advance. When discussing your answers with you learning partner, invite that person to fully explore your thinking and the logic of your answers. Be prepared to ask similarly probing questions about your partner's thinking.

Exercise 7.5

Take a moment and reflect on your understanding of statistics now. Write a few comments on the space below. What does this say about your progress in learning statistics?

____

____

____

____

____

____

8 Advanced Topics

Binomial Distribution

The simplest kind of nominal data is to obtain the proportion of a given outcome when there are only two alternatives. If we flip a coin a set number of times and consider heads to be the desired outcome (called a success) what kind of outcomes would we expect? What outcomes might we actually get? The results of multiple coin flips can be predicted based on the distribution of all possible outcomes. We have learned earlier that when we are considering the set of all possible outcomes, the proportion of a given outcome is the same as the probability of that outcome.

If we flip a coin a number of times we can calculate the proportions and hence the probabilities of various outcomes:

1. For a single toss, the possible 2 outcomes and their associated proportions and probabilities are:

    H = 1 out of 2 = ½ = .50
    T = 1 out of 2 = ½ = .50

2. For two tosses, the possible 4 outcomes and their associated proportions and probabilities are:

    2H = 1 out of 4 = ¼ = .25
    H & T (could be HT or TH) = 2 out of 4 = 2/4 = .50
    2T = 1 out of 4 = ¼ = .25

3. For three tosses, the possible 8 outcomes and their associated proportions and probabilities are:

    3H = 1 out of 8 = 1/ 8 = .125
    2H & 1T (could be HHT, HTH, or THH) = 3 out of 8 = 3/8 = .375
    1H & 2T (could be HTT, THT, or TTH) = 3 out of 8 = 3/8 = .375
    3T = 1 out of 8 = 1/8 = .125

4. For four tosses, the possible 16 outcomes and their associated proportions and probabilities are:

4H = 1 out of 16 = 1/16 = .0625
3H & 1 T (could be HHHT, HHTH, HTHH, or THHH) = 4 out of 16 = 4/16 = .25
2H & 2T (could be HHTT, HTHT, HTTH, TTHH, THTH, or THHT) = 6 out of 16 = 6/16 = .375
1H & 3T (could be HTTT, THTT, TTHT, or TTTH) = 4 out of 16 = 4/16 =. 25
4T = 1 out of 16 = 1/16 = .0625

5. For five tosses, the possible 32 outcomes and their associated proportions and probabilities are:

5H = 1 out of 32 = 1/32 = .03125
4H & 1T (could be HHHHT, HHHTH, HHTHH, HTHHH, or THHHH) = 5 out of 32 = 5/32 = .15625
3H & 2T (could be HHHTT, HHTHT, HTHHT, THHHT, THHTH, THTHH, TTHHH, HHTTH, HTHTH, or HTTHH) = 10 out of 32 = 10/32 = .3125
2H & 3T (could be TTHHH, TTHTH, THTTH, HTTTH, HTTHT, HTHTT, HHTTT, TTHHT, THTHT, or THHTT) = 10 out of 32 = 10/32 = .3125
1H & 4T (could be HTTTT, THTTT, TTHTT, TTTHT, or TTTTH) = 5 out of 32 = 5/32 = .15625
5T = 1 out of 32 = 1/32 = .03125

6. Finally for six tosses, the possible 64 outcomes and their associated proportions and probabilities are:

6H = 1 out of 64 = 1/64 = .015625
5H & 1T (could be HHHHHT, HHHHTH, HHHTHH, HHTHHH, HTHHHH, or THHHHH) = 6 out of 64 = 6/64 = .09375
4H & 2T (could be 13 different combinations) = 15 out of 64 = 15/64 = .234375
3H & 3T (could be 24 different combinations) = 20 out of 64 = 20/64 = .3125
2H & 4T (could be 13 different combinations) = 15 out of 64 = 15/64 = .234375

1H & 5T (could be 6 different combinations) = 6 out of 64 = 6/64 = .093756T = 1 out of 64 = 1/64 = .015625

If we look at the distributions created from these results (called Binomial distributions), we see that the larger the number of tosses, the more the distribution begins to look like the normal distribution. (The corresponding normal curve has been imposed over the plots for the six outcomes above to show how these plots approach the normal

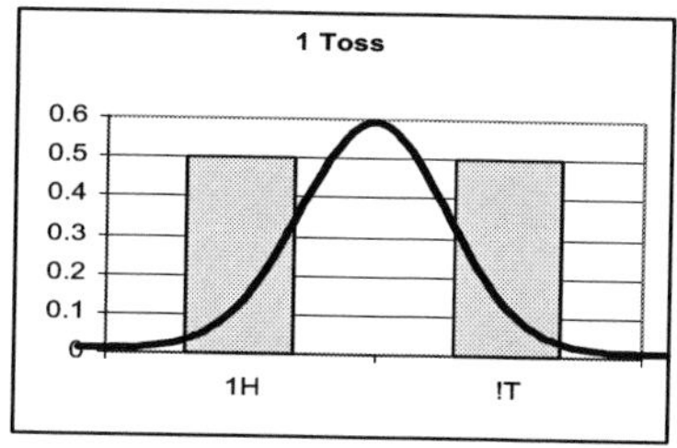

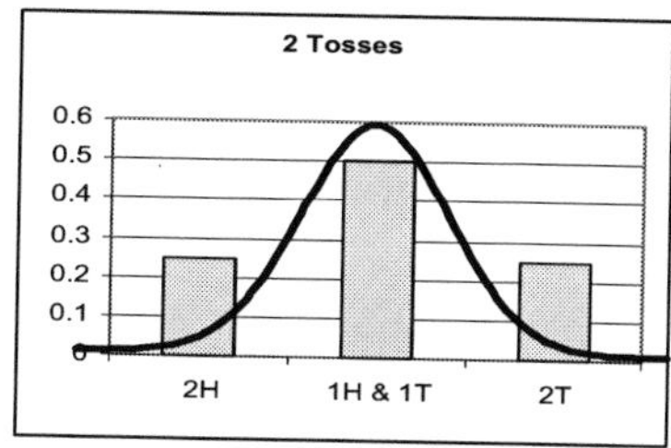

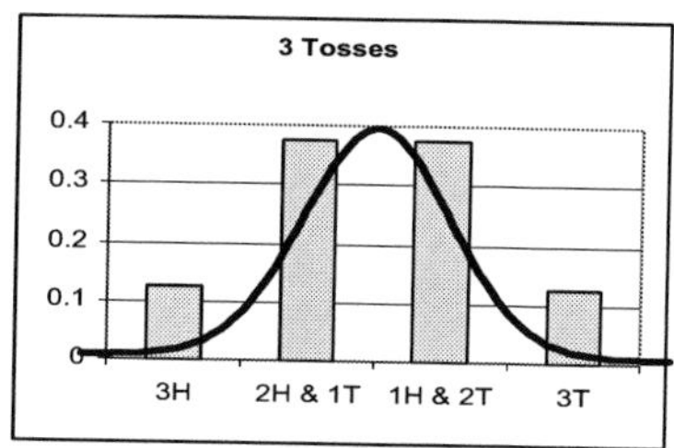

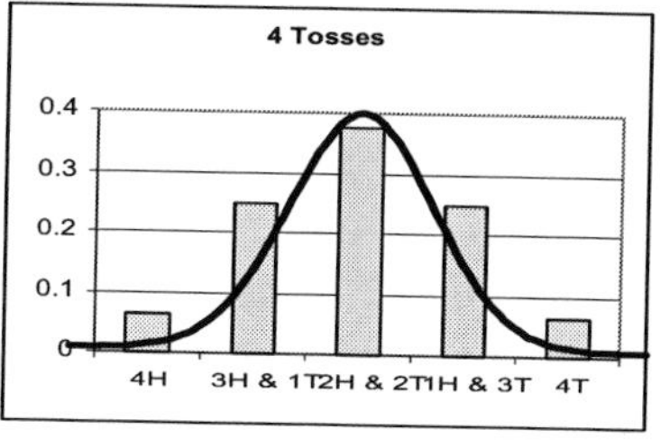

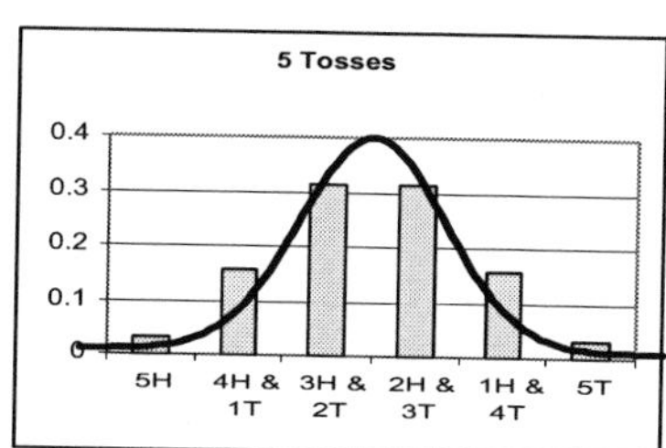

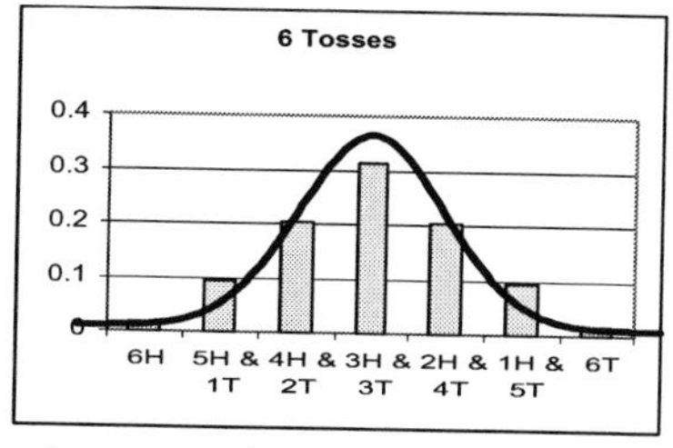

distribution as the number of coin tosses increases.) Note that the coin toss results are nominal data; the counts of results are a proportion for each outcome. As we focus on a proportion in a given category (number of successes versus number of non-successes) the probability distribution begins to look more and more like the normal curve as the number of events (coin tosses) increases. If we increase the number of tosses to twenty, the number of unique outcomes increases to 1,048,576, and the number of head/tail combinations increases to 21. The probability

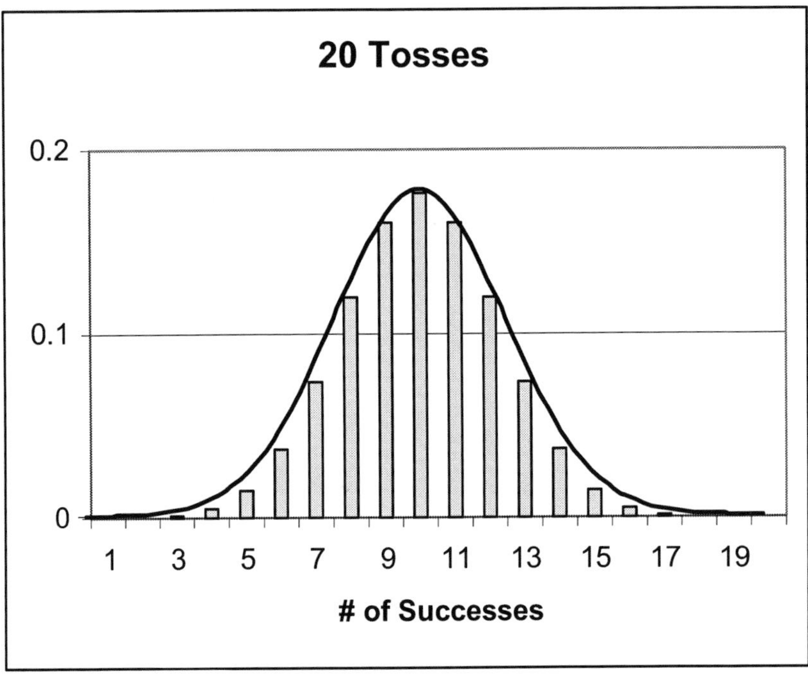

distribution of these 21 combinations is almost indistinguishable from the normal curve superimposed on it, as illustrated above. (To simplify the distribution only the number of heads, called a "success" is displayed) So if simple probability events like coin tosses can come to be represented by the normal curve, it may be possible to represent such simple outcomes in terms of an average value and a standard deviation. Another way to think of the drawing together of these concepts is: a probability is a probability is a probability. The probability of a combination of heads and tails from a given number of coin tosses is a result of the number of ways those results can be achieved as a proportion to the total number of probable outcomes. When the number of outcomes of simple probabilities gets large enough, the probability distribution merges to the normal curve and its values and properties apply.

Simple Proportions

Suppose we toss a coin 10 times and achieve a result of 5 H and 5 T. If we focus on one outcome (e.g. heads) and call it a "success", then the proportion of our successful outcomes is symbolized by the letter p (for proportion) with a carat (^) over it $\hat{p}$. This is the symbol (pronounced "p-carat" or "p-hat") for a sample proportion and the corresponding symbol is p for the true proportion of a population. (Since true proportion for a population is the same as the probability, p can also stand for the probability of the outcome.) The distributions in the examples above are the probability distributions for the possible outcomes for coin tosses when the number of tosses is from 1 to 6 or 20. In all cases the sample distributions center on the "true" value for the population which is .5 or 50% heads. Thus the center (or average) outcome for the various sample proportions $\hat{p}$ (symbolized $\bar{x}(\hat{p})$) is 0.5 which is the same as the "true" value for the population (of all possible outcomes of coin tosses). Just like interval and ratio data where $\bar{x}$ (the sample mean) is said to be an unbiased estimator (good guess for) of μ (the population mean), the sample proportion $\hat{p}$ is said to be an unbiased estimator of p the population proportion. How good an estimator it is will depend upon the sample size, n. Another way to say this is the average of all sample proportions is the true population probability:

$$\bar{x}(\hat{p}) = p$$

So if our normal curve, superimposed on the binomial distribution, centers on the true population proportion p, what is the spread or standard deviation (s) associated with this normal curve in terms of the binomial distribution? Before presenting the equation for the value of s, consider the relationship of the normal curves presented previously for n = 1 to 6 and 20 tosses. When n is small, the normal curve does a very poor job of representing the binomial distribution. As n increases the binomial distribution begins to look more like the normal curve. The spread to this sampling distribution normal curve, expressed in the binomial probabilities is:

$$s(\hat{p}) = \sqrt{\frac{p(1-p)}{n}}$$

Or in words: the standard deviation (the spread) of sampling distribution of proportions for a given sample size (n), is the square root of the true population proportion (or probability) times one minus the true proportion, divided by the sample size (n). Let's compare this to the equation for the spread of sample data for interval or ratio data.

$$s_{\overline{X}} = \frac{\sigma}{\sqrt{n}}$$

If we put these two equations side by side, we see:

$$s(\hat{p}) = \sqrt{\frac{p(1-p)}{n}} \text{ and } s_{\overline{X}} = \frac{\sigma}{\sqrt{n}}$$

and we can see that just like with interval and ratio data, the spread for sample averages is narrower the larger the sample size. How large a sample is needed to ensure the binomial distribution is effectively represented by the normal curve? Both the actual count of successes (equal to proportion times the sample size, $\hat{p}$ x n = number of successes) and the count of non-successes (equal to (1 - $\hat{p}$) x n = number of non-successes) must be 10 or greater. By this criteria, the binomial of 20 coin tosses, n = 20 would be necessary for the normal curve to replace the binomial distribution. Just as is the case with interval or ratio data where the true population values may not be known, if we do not know the value of p for the population, we substitute the sample values:

$$s(\hat{p}) = \sqrt{\frac{\hat{p}\ (1-\hat{p}\ )}{n}}$$

How can we make use of this relationship? Let's use the binomial distribution to understand some probable outcomes from coin tosses. The binomial distribution is defined from simple probabilities, and this definition has been used to prepare the table of binomial probabilities in Appendix 3. This table displays the probabilities for a given number of "successes" out of a total on "n" measurements; for population proportions (true probabilities) "p" of .01, .02, .03, .04, .05, .06, .07, .08, .09, .10, .15, .20, .25, .30, .35, .40, .45, and .50. The values of "n" listed in the table are 1, 2, 3, 4, 5, 6, 7, 8, 9, 10, 11, 12, 15, and 20. Numbers of successes correspondingly range from 0 to "n".

How do you use the table? Suppose that you want to determine the likelihood (probability) of obtaining 5 heads from seven coin tosses (n = 7, # of successes = 5, p = 0.5). You would enter the table for the value of n = 7, for 5 successes, and go to the column under p = .50. There you would find a value of 0.1641, which is the probability of obtaining five heads from 7 coin tosses. What would be the probability of obtaining 2 heads from 7 coin tosses? Now enter the same table for 2 successes, and find a value of 0.1641, the same value. Why is this? The probability of obtaining 2 heads (= 5 tails) is the same as obtaining 5 heads (= 2 tails), because the probability of success and failure is the same, 0.5. If our coin were "loaded" so as to decrease the probability of heads to 0.45 how would we calculate these two outcomes? Now we would enter the table for the two levels of successes at a probability of 0.45. In this case the probability of 2 heads would be 0.2140 while that of 5 heads would be 0.1172, quite a bit lower.

Exercise 8.1

A baseball player has a lifetime batting average of .300, in the new season he will come to bat 100 times. What is his expected number of hits, and what will be the standard deviation for that number of hits?

The expected outcome $\hat{p}$ would be the same as p = 0.3 (lifetime average is the population), so the number of expected hits is equal to ($\hat{p}$ x n) = 0.3 x 100 =30.

$\bar{x}\,(\hat{p}) = p$

$= .300$ *(as a proportion),*

$= (.300)(100)$

= 30 (as number of hits)

The standard deviation is:

$$s(\hat{p}) = \sqrt{\frac{p(1-p)}{n}}$$

$$= \sqrt{\frac{.3(.7)}{100}}$$

$$= \sqrt{\frac{.21}{100}}$$

$$= \sqrt{.0021}$$

= .046 (as a proportion), and for the 100 times at bat, the standard deviation is (.046)(100)

= 4.6 (as number of hits).

Exercise 8.2

What would be the 95% confidence interval for the batting average for the above player this season?

Begin by reviewing what we know: for this baseball player, the expected number of hits is 30 with a standard deviation of 4.6 hits. Assume the average hits will be 30 (reasonable since $\hat{p}$ is an unbiased estimator of p), and a normal distribution will represent the distribution of hits with an s = 4.6. From the normal curve, we recall that a 95% confidence interval corresponds to ±2 standard deviations. To create the confidence interval we remember that:

$$95\%\ C.I. = \bar{x} \pm 2\ s$$

$$95\%\ C.I. = 30 \pm (2)(4.6)$$

$$95\%\ C.I. = 30 \pm 9.2$$

So: 95% C.I. = approximately 21 to 39 hits for 100 at bats this season. We would say "there is a 95% probability that this batter will get from 21 to 39 hits out of 100 at-bats this season".

So we can prepare probability statements for nominal (counting) data using the normal curve and its associated relationships, when the number of success or non-successes for the counts exceeds 10. If the number of successes or non-successes is less than 10, then the values and probabilities from the binomial distribution have to be used.

How do we use the binomial table when $p > 0.50$? This binomial table in appendix 3 only covers values of p from 0.01 to 0.50. Just like the Z table which only covers half of the normal curve, we use symmetry properties of the binomial distribution for values of $p > 0.50$. If the p or $\hat{p}$ for success is greater than 0.5, do the calculations substituting the non-success $(1 - p)$ data. The results will be the value for non-successes and the value for successes can be determined by subtracting the non-successes value from 1. Let's demonstrate with an exercise:

Exercise 8.3

Joe NBA is a basketball player with a lifetime average free throw success rate of 0.75. During the upcoming season he is expected to attempt approximately 400 free throws. a.) What is the probability that he will obtain exactly 15 successes in his first 20 free throw attempts?

b.) What is the 95% confidence interval for his number of successes for this whole season?

The question for part a.) is: what probability is associated with 15 successes out of 20 attempts, where there is a .75 probability of success? If we go to the binomial table, we see that it addresses 15 successes in 20 attempts, but only to a maximum probability of 0.5. But we need a value of this for p = .75! Here's how we do this. Notice that 15 successes out of 20 equates to 5 non-successes out of 20. If the probability associated with success is 0.75, then the probability associated with non-success is 0.25. (1 – p = probability of non-success.) So this question could be reformulated to ask: what is the probability of 5 successes out of 20 when the probability of success is .25? (We have just flopped to calling a non-success a success) From the table we find the following value: for n = 20, 5 successes at p = .25, the probability is 0.2023. Thus there is a 20.23% probability that Joe will get exactly 15 successful free throws during his first 20 tries.

For part b.) we need to determine the value of $\hat{p}$ *and s(*$\hat{p}$*) to predict the 95% C.I. as previously. From the value of p = .75 for Joe's lifetime (population) average, we can assume* $\hat{p}$ *= .75 for this season (sample). Thus the number of successful free throws will be (.75)(400) = 300. To predict the 95% C.I. we need the value of s(*$\hat{p}$*).*

$$s(\hat{p}) = \sqrt{\frac{p(1-p)}{n}}$$

$$s(\hat{p}) = \sqrt{\frac{.75(.25)}{400}}$$

$$s(\hat{p}) = \sqrt{\frac{.1875}{400}}$$

$$s(\hat{p}) = \sqrt{.00046875}$$

$s(\hat{p})$ = *.022, and for 400 free throw attempts, s = (.022)(400) = 8.67 successes.*

From the normal curve, a 95% confidence interval corresponds to a Z value of 2. (A purist would point out that through the Z table it is really 1.96 standard deviations that correspond to 95% probability, rather than 2 standard deviations. The difference corresponds to an "error" of 2% in the calculated results.) To create the confidence interval we remember that:

$$95\%\ C.I. = \bar{x} \pm Z_{.95}\,(s)$$

$$95\%\ C.I. = 300 \pm (2)(8.67)$$

95% C.I. = 300 ± 17.3 = approximately 283 to 317 successful free throws this year.

The Law of Large Numbers

When we start to perform coin tosses, we can begin to get results that seem to depart from the expected outcome of 50/50 for the ratio of heads to tails. As we have learned previously, the departure from the expected

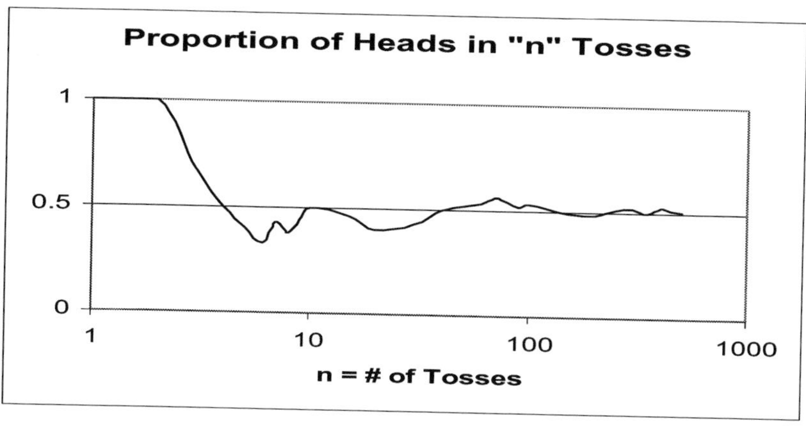

value can be rather large when we are considering the results from a relatively small number of tosses. As the number of tosses increases, the results for the cumulative tosses, $\hat{p}$, will begin to get closer and closer to the population true proportion, p. Mathematicians first discussed this relationship, called the law of large numbers, in the late 1600's. In order for the law of large numbers to apply, the phenomenon being tracked must be truly random with each outcome independent of the previous outcomes. The law of large numbers is the basis for the success of both gambling casinos and insurance companies. When many people are insured, though the fate of any insured individual is random and unpredictable, the fate of the total pool is subject to the law of large numbers and is predictable. Rates and payoffs are adjusted to assure the insurance company remains viable and profitable.

Central Limit Theorem

When a sample is obtained from a population and the average, $\bar{x}$, for that sample is calculated, you are developing the basis for creating a distribution of the sample averages. If many samples of the same size are taken, and their averages are calculated and plotted, the result is called a sampling distribution of the sample averages. What kinds of sampling distributions can we expect if they are created from very different population distributions? If a sampling distribution is created from a normal distribution, its shape will also be normal. This makes sense; an initial expectation would be that the sampling distribution of $\bar{x}$ would mimic the population distribution. Is this true for sampling distributions from populations that are non-normal? Computer simulations[4] have been created for sampling distributions from highly non-normal populations to demonstrate the effect of sample size on resultant sampling distributions. As illustrated on the next page, even highly non-normal distributions yield normal sampling distributions when the sample size (n) becomes sufficiently large. How large a sample is required depends upon the shape of the population distribution. However, as illustrated, even very non-normal distributions yield normal sampling distributions with sample sizes approaching 30. The only requirement for the central limit theorem to apply to sampling distributions for any population is that the population has a calculable value of μ and σ.

In the illustration following, the very non-normal parabolic, exponential and square wave population distributions all produce a normal sampling distribution of $\bar{x}$ by the time the sample size equals 30. (Remember for the binomial distribution, the sampling distribution of $\hat{p}$ had become normal when n = 20.)

Another aspect of a sampling distribution is the fact that averages of many large samples (when n is large) show a much narrower sampling distribution (have less spread) than is the case for averages of many smaller samples (n is smaller). We can create a symbol for the spread (standard deviation) of the distribution of (many) averages for a given sample size (n). The symbol ($s_{\bar{x}}$) combines the symbol of sample standard deviation (s) with that of the sample average ($\bar{x}$) and is pronounced "s – sub-x-bar". This standard deviation of a sampling distribution of means ($s_{\bar{x}}$) relates to the spread of the original population

---

[4] Kiemele, M. J., S. R. Schmidt, and R. J. Berdine. (1999). *Basic Statistics Tools for Continuous Improvement (4th ed.)*. Colorado Springs, CO: Air Academy Press [ISBN: 1-880156-06-7], page 5-17 to 5-18.

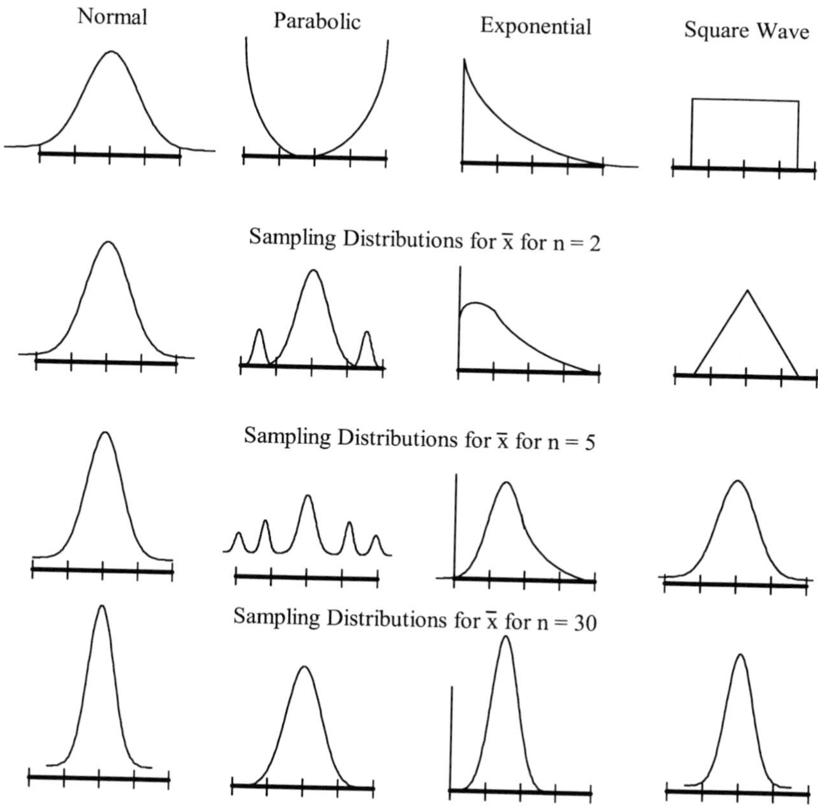

Illustration of the Central Limit Theorem

(the larger the population standard deviation σ, the larger the spread of sample averages) and to the sample size (the larger the sample size n, for the many averages, the narrower the averages' sampling distribution). This relationship is expressed as the standard deviation of the mean:

$$s_{\overline{X}} = \frac{\sigma}{\sqrt{n}}.$$

Thus the spread of the curves for n = 30 is much narrower than for n = 5.

To demonstrate the relationship of the standard deviation of the mean (also known as standard error) for various sampling distributions, the author performed a computer simulation for a very simple population. The population is like the square wave population above. It consists of the values: 1, 2, 3, 4, and 5. As illustrated on the following page, this population has an average of 3.0 and a standard deviation of 1.41. For the first simulation all 25 possible samples of size n = 2 (sampling with replacement) were created and their averages and standard deviations calculated (Standard deviations were calculated both using n only as if a population standard deviation, and using n –1 as for samples.) The 25 averages were then plotted, and the standard deviation of the 25 averages was calculated. This measured standard deviation of the 25 means is compared to the predicted standard deviation based on the formula: $s_{\overline{X}} = \frac{\sigma}{\sqrt{n}}$. The resultant data are shown. Then all 125 possible samples of size n = 3, were created and the corresponding values developed. Finally all 625 possible samples of size n = 4 were also created and the same data developed. Some of the more difficult issues to comprehend about samples are well illustrated by the data.

Use of (n – 1) to Calculate Standard Deviations for Samples

When the value of (n) is used to calculate the sample standard deviations for the three sets of samples, each pooled standard deviation (square root of the average of the variances for the 25, 125 and 625 sample sets) underestimates the true population standard deviation. The amount of the underestimation is exactly compensated when the value of (n – 1) is used to instead calculate each sample standard deviation and used to create the pooled value. The 25 samples of size n = 2 produced a pooled standard deviation of 1.00 when calculated using (n), the same 25 produced the true pooled standard deviation of 1.41 when calculated using (n - 1). Similar results were true for the samples of sizes n = 3 and n = 4 as shown beside each of the sample average distributions.

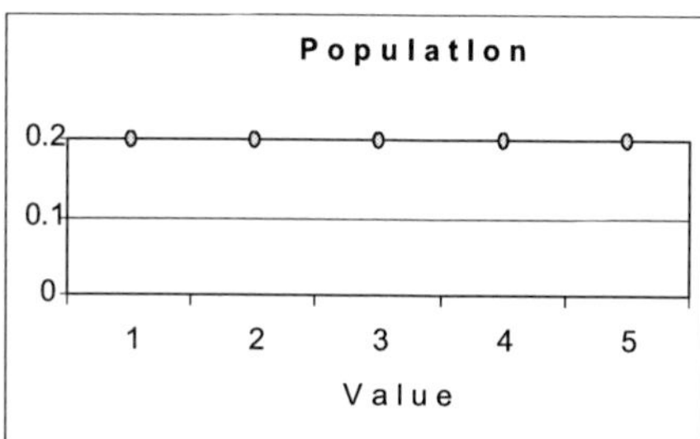

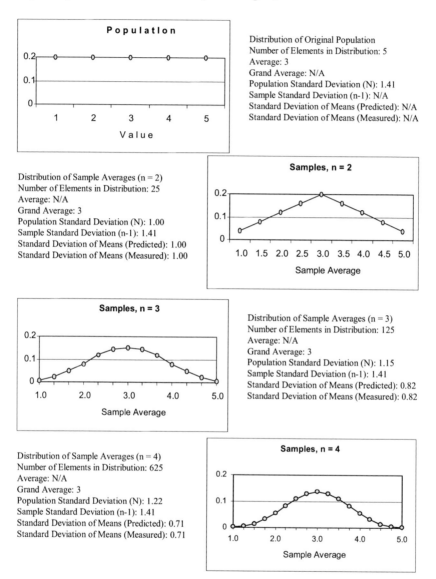

Distribution of Original Population
Number of Elements in Distribution: 5
Average: 3
Grand Average: N/A
Population Standard Deviation (N): 1.41
Sample Standard Deviation (n-1): N/A
Standard Deviation of Means (Predicted): N/A
Standard Deviation of Means (Measured): N/A

Distribution of Sample Averages (n = 2)
Number of Elements in Distribution: 25
Average: N/A
Grand Average: 3
Population Standard Deviation (N): 1.00
Sample Standard Deviation (n-1): 1.41
Standard Deviation of Means (Predicted): 1.00
Standard Deviation of Means (Measured): 1.00

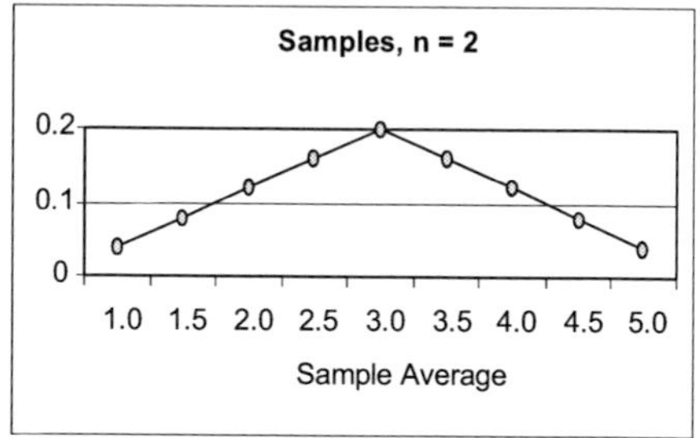

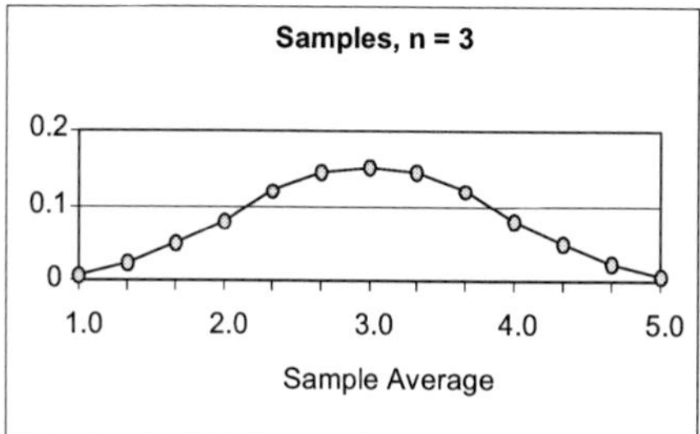

Distribution of Sample Averages (n = 3)
Number of Elements in Distribution: 125
Average: N/A
Grand Average: 3
Population Standard Deviation (N): 1.15
Sample Standard Deviation (n-1): 1.41
Standard Deviation of Means (Predicted): 0.82
Standard Deviation of Means (Measured): 0.82

Distribution of Sample Averages (n = 4)
Number of Elements in Distribution: 625
Average: N/A
Grand Average: 3
Population Standard Deviation (N): 1.22
Sample Standard Deviation (n-1): 1.41
Standard Deviation of Means (Predicted): 0.71
Standard Deviation of Means (Measured): 0.71

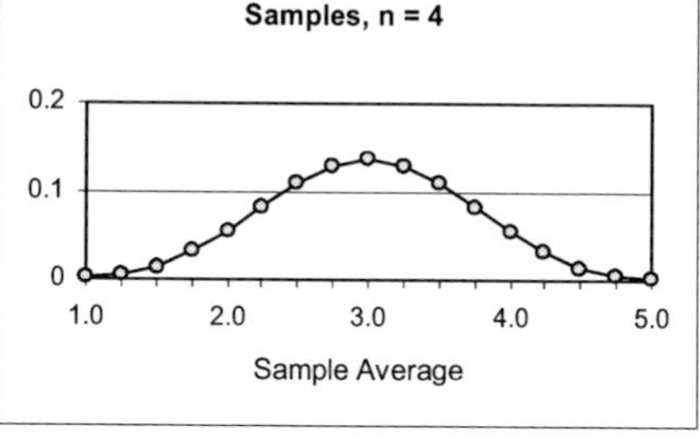

Validation of the Standard Deviation of the Mean Calculation

The relationship of the standard deviation of the distribution of means for a given sample size: $s_{\overline{X}} = \frac{\sigma}{\sqrt{n}}$ is derived from the Central Limit Theorem. The simulations demonstrate the validity of this relationship. When all possible samples of the given sample size are created and the standard deviation of those samples' average values is calculated, the results exactly match the calculation above based on the population standard deviation. In the most extreme case, samples of size n = 4, the 625 sample averages were used to calculate their standard deviation. The result was a value of 0.707107, rounded to 0.71 in the preceding results. The population standard deviation for the five element population is $\sqrt{2}$ = 1.414214. When this value is used to calculate the predicted value of the standard deviation of means for sample size n = 4, we get: $s_{\overline{X}} = \frac{\sigma}{\sqrt{n}} = \frac{1.414214}{\sqrt{4}} = \frac{1.414214}{2} = 0.707107$. This number is exactly the same as the measured value for the 625 samples developed. The same result was true for samples of size n = 2 and n = 3.

Exercise 8.4

If a population has $\mu = 100$ and $\sigma = 25$, what would be the standard deviations for sampling distributions of means for samples of sample sizes: $n = 15$, $n = 50$, $n = 100$, $n = 250$, and $n = 1500$? Sketch on the following page the five sampling curves you would obtain if you took all possible samples of each sample size. (Note the change in scale from the population curve.)

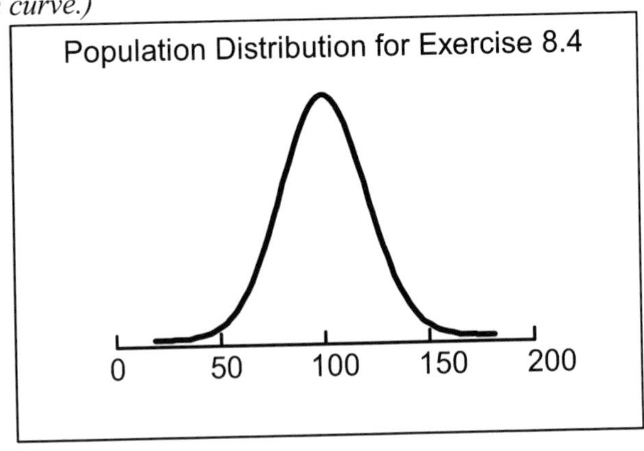

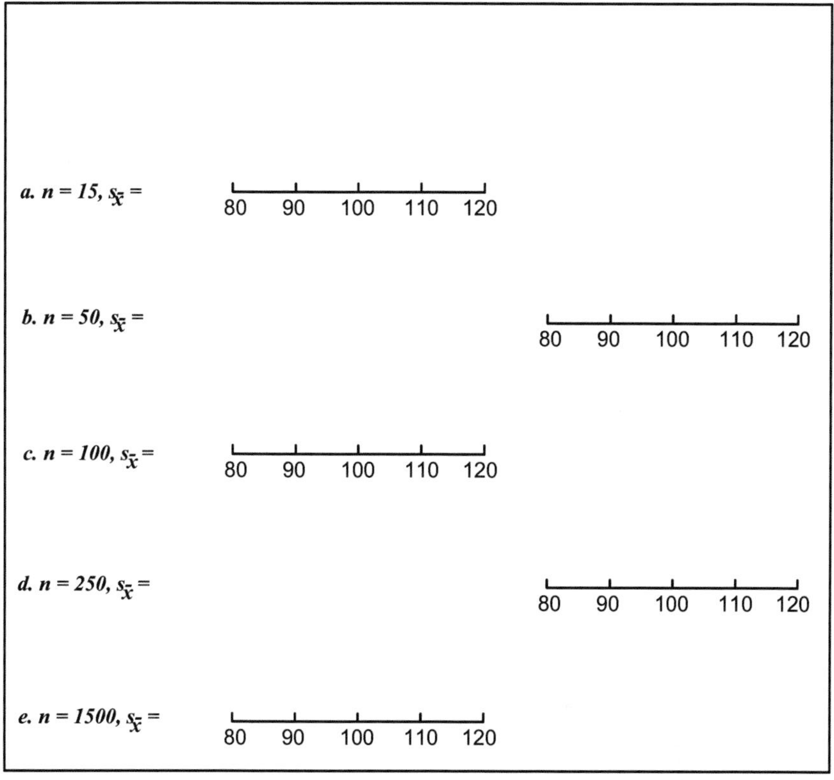

The answers appear below. The calculations are simple, the comprehension of their meaning is very important. First the calculations:

a. $n = 15, \ s_{\bar{x}} = \frac{\sigma}{\sqrt{n}} = \frac{25}{\sqrt{15}} = \frac{25}{3.87} = 6.46$

b. $n = 50,\ s_{\overline{X}} = \dfrac{\sigma}{\sqrt{n}} = \dfrac{25}{\sqrt{50}} = \dfrac{25}{7.07} = 3.54$

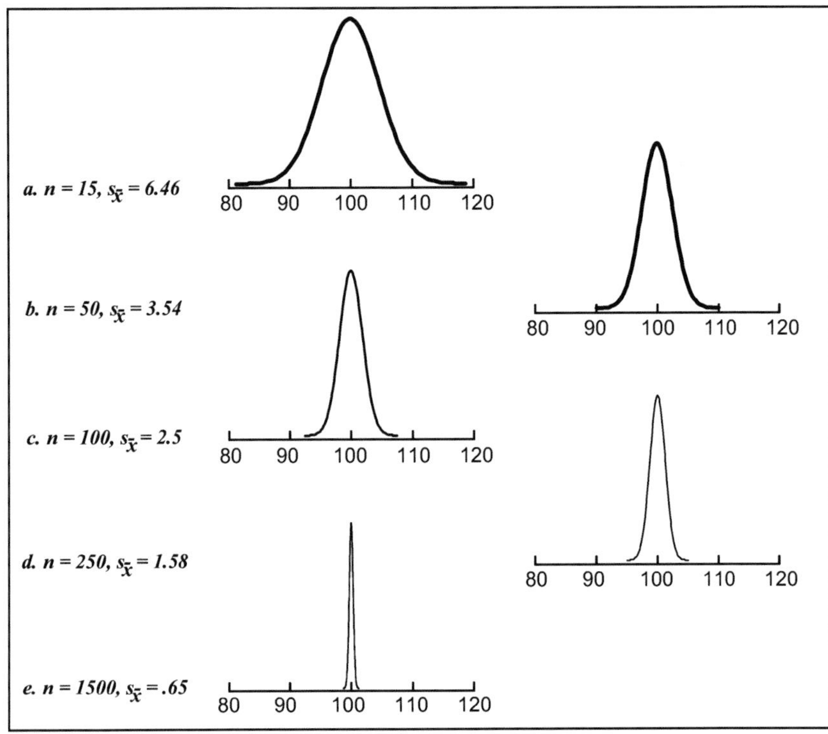

c. $n = 100,\ s_{\overline{X}} = \dfrac{\sigma}{\sqrt{n}} = \dfrac{25}{\sqrt{100}} = \dfrac{25}{10} = 2.5$

d. $n = 250,\ s_{\overline{X}} = \dfrac{\sigma}{\sqrt{n}} = \dfrac{25}{\sqrt{250}} = \dfrac{25}{15.81} = 1.58$

e. $n = 1500,\ s_{\overline{X}} = \dfrac{\sigma}{\sqrt{n}} = \dfrac{25}{\sqrt{1500}} = \dfrac{25}{38.7} = .65$

The resulting normal curves are illustrated above. Notice that the curves for n = 250 and 1500 are nearly straight lines. Using these two sampling distributions, the resulting 95% confidence intervals would be very narrow. In other words very large sample sizes allow us to predict population values very precisely.

9 Performing Statistical Tests

The Basis of Statistical Tests

We have already defined the Null and Research hypotheses as they relate to statistical tests. Fundamentally, every statistical test is asking and answering the following question: What is the probability that these two (or more) sets of results come from the same source (i.e., a distribution, a population, etc.), and given that probability, what do we conclude about the Null hypothesis? So lets spend some time looking at these inter-related questions and how statistical tests deal with them.

Every statistical test is looking at two (or more) sets of data in order to answer the question: are they the same (accepting H_0) or are they different (rejecting H_0 and accepting H_1)?

1. Suppose we start this inquiry with the following data:
   Result 1 = 49.7
   Result 2 = 49.7
   Measure of difference = 0.0

H_0 states: there is no difference between these two results. The beginning of our evaluation to create a test is to ask: what is the probability value that we can attribute to H_0? In this circumstance where the measure of difference is zero we have an essentially 100% probability associated with H_0. We therefore would be wise to accept H_0 because it is so highly probable to be correct.

2. We then continue the inquiry with the following data:
   Result 1 = 49.7
   Result 2 = 49.8
   Measure of difference = 0.1

Again we perform an evaluation of the probability value that we can attribute to H_0. In this case suppose we find that there is a 95% probability associated with H_0. Again we are wise to accept H_0 given this level of probability that H_0 is correct.

We could continue these examples by looking at more data showing increasingly larger differences between the first and second result. As these measures of difference increase, the associated probabilities of H_0 being correct are correspondingly decreasing. As we progress in this manner, we continue to ask the question: is this measure of difference yet large enough to reject H_0? The basis for answering this continuing question is to determine the corresponding probability that this difference is reflective of H_0 being correct.

Each different statistical test has a different basis for calculating the probability value associated with H_0. The probability values depend upon: the type of data (level of scaling), the amount of data available (degrees of freedom), the nature of the underlying distribution (normal, skewed, etc.), whether data are paired or matched, etc. The rest of this book provides the details of six common statistical tests used for the evaluation of these probabilities associated with measures of differences.

3. Suppose in the progression above, the next set of results were:

   Result 1 = 49.7
   Result 2 = 50.3
   Measure of difference = 0.6

We now recalculate the probabilities that these results represent H_0 being correct and develop a 75% probability that this measure of difference does reflect H_0. What should we conclude? We would still conclude that we accept H_0. If we did not accept, but rather rejected H_0, we have a 75% probability we have made a wrong decision, i.e. rejecting H_0 is wrong 3 out of 4 times.

So how big a measure of difference and correspondingly how small a probability that the difference reflects H_0 do we need to reject H_0? The preceding probability values of essentially 100%, 95%, and 75% were not small enough to reject H_0. How about 50%? How about 25%, 20%, 15%, 10%, 5% or 1%? The larger the (relative) measure of difference is between two or more sets of results, the smaller the related probability of the "truth" of the null hypothesis. The method of assigning a probability value to a measure of difference is dictated by all the factors listed previously. How small the probability that H_0 is correct has to be before we decide to reject H_0 is determined by the person performing the test.

For each test a critical statistical value is decided in advance of performing the test. If the calculated value of probability for the measure of difference is smaller than that value, H_0 is rejected. If the calculated probability is larger than that critical value, H_0 is accepted. The critical value, which we pre-determine for a given test is called the level of significance and is represented by the symbol α. For test of difference related to the physical sciences values of α may be as small as 1% or even 0.1%. For the behavioral sciences, larger values of α are usually used: typically 5% and even occasionally 10%.

In all cases the progression of the statistical test is similar:

a. Create the statement of the null and research hypotheses, H_0 and H_1.
b. Determine the applicable level of significance, α.
c. Calculate an appropriate measure of difference for the data sets being compared. Depending on the types of data and the test being performed, this may be a true difference measure, a ratio of values, or another expression of differences.
d. Using the appropriate calculations or statistical tables (see following) convert the measure of difference to a probability related to accepting H_0.
e. If the probability of H_0 being correct equals or exceeds the selected level of significance α, accept H_0; if the probability of H_0 being correct is smaller than α, reject H_0.

A Word About the Use of Tables

As we begin the detailed practice of performing statistical tests, it is necessary to become more familiar with the use of statistical tables. The calculations that underlie the performance of statistical tests are both complex and unique to a given test circumstance. Long before the computational capabilities of computers were widely available, statistical tables were laboriously prepared using hand calculations. Each test application is often unique as to the number of data, desired level of significance, type of data and other circumstances that require a unique solution value. By developing general tables, statisticians were able to accommodate these variations with a wide-ranging set of numerical solutions applicable to the most common test circumstances.

Anyone with a home personal computer with a good spreadsheet program now has the capability to rapidly and accurately perform most statistical tests. Details of specifically which tests can be performed reside in the documentation for the specific software. This text does not take on the challenge of addressing the myriad available personal computing options. Instead the appropriate statistical tables are re-created and utilized in the tests presented.

How to use a specific table depends upon the specific test (and sometimes details of its application) to be used. It is usually necessary that the details of the test and the specifics of the unique application be well understood. Clearly you must know how much data you have, its associated degrees of freedom (usually related to sample size as n - 1), to what level of significance (α) the test will be performed, and other information specific to the test application. Only then can you go to the specific row and column in the appropriate table to acquire the necessary test comparison value. How that value is compared to the test statistic you have calculated will also depend upon the type of test being performed.

Chi-Square Test of Significance

The Chi-Square test is one of the most universal tests of significance for statistical applications. It is usable with all levels of scaling: nominal, ordinal, interval or ratio. The fundamental question being tested with the chi-square test is this: does a set of sample outcomes for frequencies or "counts" differ significantly from an "expected" set of outcomes? The expected values can arise from population information (previously known values), from a theoretically expected distribution of the values, or from expected values based on the null hypothesis. All three sources will be demonstrated in exercises.

The chi-square is a non-parametric test that relies on probability relationships for its power and does not require either a particular type of data or a particular type of data distribution. In this aspect it is less powerful than parametric tests would be. (Power of a test relates to the amount of data necessary to reach a conclusion at a pre-determined value of α. More powerful tests require less data to reach a conclusion for the given level of significance than less powerful tests.)

The method for constructing the chi-square (symbolized χ^2) test value relates to adding together the squared differences between Observed and Expected values (usually expressed as counts), divided by the expected values. This will be shown in the equation below after we discuss the construction of the value of χ^2. Chi-square is fundamentally a difference measure, and just as we calculate standard deviations by squaring differences, the chi-square begins with squared differences. It then proceeds to relate those squared differences to the Expected value, since the Expected value represents the theoretical or correct outcome. Finally the value of chi-square results from adding up these differences for all outcomes being evaluated.

Requirements for the Chi-Square Test[5]

1. Nominal Data
2. One or more groups (Columns in a matrix)
3. One or more categories (Rows in a matrix)
4. Independent observations
5. Adequate sample size:
   a. Expected frequencies should be sufficiently large for 50% of categories with values of five or larger
   b. When there are more than two categories, no category should have values less than two
6. Simple random sample
7. Data in frequency (counts) form
8. All observations must be used
9. Two-tail test only (no one-tail test is possible)

Chi-Square can be used when we generate the expected data from the observed data. The following exercise shows how the null hypothesis can be used to generate a set of expected values from the corresponding set of observed data. The steps used in this example will be used in many of the applications of Chi-Square.

---

[5] Format and content based on Sharp, Vicki F. (1979), *Statistics for the Social Sciences*. Boston, MA: Little Brown and Company (Library of Congress Catalog Card No. 78-70849), page 182.

Exercise 9.1

A study of enrollees in a Ph.D. program examined their status after six years to understand whether the gender of the students affected the status. The data was:

Status	*Male*	*Female*
Graduated	*423*	*98*
Still-enrolled	*134*	*33*
Dropped-out	*238*	*98*

Prepare a set of Expected values for the above data based on the null hypothesis that there is no difference in the relative rates between males and females. Perform a test at $\alpha = 0.05$.

In order to determine the Expected values, we will use the assumptions inherent in the null hypothesis. If there is no difference in the experience of males and females, their rates of completion, still enrolled and dropping-out should be proportional to their proportion in the overall sample. To prepare the expected values, we have to calculate the sums for the above table both across the rows and down the columns. These sums will be used to calculate the overall proportions of females and males and separate proportions across the three categories. These proportions will then be used to calculate the expected values. It helps to embed these calculations in an expansion of the above table. The first step is to calculate the totals for all columns and rows:

Status	*Male*	*Female*	$\underline{\Sigma}$
Graduated	*423*	*98*	*521*
Still-enrolled	*134*	*33*	*167*
Dropped-out	*238*	*98*	*336*
$\underline{\Sigma}$	*795*	*229*	*1024*

Now we calculate the Expected proportions for each category. Males as a proportion of the total $= \dfrac{795}{1024} = 0.776$*; Females as a proportion of* *the total* $= \frac{229}{1024} = 0.224$*. The* *Graduated* *category as a proportion of the total* $= \frac{521}{1024} = 0.509$*; the* *Still-Enrolled* *category as a proportion of the total* $= \frac{167}{1024} = 0.163$*; and the* *Dropped-Out* *category as a proportion of the total* $= \frac{336}{1024} = 0.328$*. These proportions are now used to calculate the Expected counts for each cell by multiplying the total (1024) times the two proportions (from the row and column values) that affect each cell:*

Expected Values

Status	*Male (0.776)*	*Females (0.224)*
Completed (0.509)	*(1024)(.509)(.776) = 404.5*	*(1024)(.509)(.224) = 116.8*
Still-enrolled (0.163)	*(1024)(.163)(.776) = 129.5*	*(1024)(.163)(.224) = 37.4*
Dropped-out (0.328)	*(1024)(.328)(.776) = 260.6*	*(1024)(.328)(.224) = 75.2*

Calculate as follows: $\chi^2 = \sum \frac{(\text{Observed - Expected})^2}{\text{Expected}}$

$$= \frac{(423-404.5)^2}{404.5} + \frac{(134-129.5)^2}{129.5} + \frac{(238-260.6)^2}{260.6} + \frac{(98-116.8)^2}{116.8} + \frac{(33-37.4)^2}{37.4} + \frac{(98-75.2)^2}{75.2},$$

$$= \frac{(18.5)^2}{404.5} + \frac{(4.5)^2}{129.5} + \frac{(-22.6)^2}{260.6} + \frac{(18.8)^2}{116.8} + \frac{(-4.4)^2}{37.4} + \frac{(22.8)^2}{75.2},$$

$= .85 + .16 + 1.96 + 3.03 + .52 + 6.91$

$= 13.43$

Perform the χ^2 evaluation as follows: for a matrix with two or more columns and rows, the degrees of freedom = (# of rows -1) times (# of columns -1), so for this application, degrees of freedom = (3-1)(2-1) = 2 x 1 = 2. We have already pre-determined the level of significance, α = .05. The critical value of χ^2 from the table (Appendix 4) for these values is 5.991. Since our calculated χ^2 of 13.43 exceeds this value we reject the null hypothesis and can accept the alternate hypothesis (that there is a difference in the status of females after six years).

We can also use the χ^2 test to verify whether data is consistent with a theoretical distribution. The theoretical distribution is used to calculate the Expected values for the data that are then compared with the Observed values.

Exercise 9.2
Reggie Jackson during regular seasons was at bat 9864 times and had 2584 hits. During his participation in World Series games, he was at bat 98 times and had 35 hits. Use χ^2 to test the null hypothesis, H_0: there is no difference between Reggie's performance in normal season play and World Series play.

The formula for the chi-square statistic is:

$$\chi^2 = \sum \frac{(\text{Observed - Expected})^2}{\text{Expected}}$$

For this exercise, the observed value is 35 hits out of 98 opportunities for a ratio of $\frac{35}{98}$ = 0.357. The normal season data was 2584 hits for 9864 times at bat resulting in a ratio of $\frac{2584}{9864}$ = 0.262. For the null hypothesis to be correct, we would expect World Series performance to be the same as the normal season. Thus we combine the total data to create the theoretical outcomes. The total at bats then would be 9864 + 98 = 9962, with total hits of 2584 + 35 = 2619. The resultant overall batting average would be $\frac{2619}{9962}$ = 0.263. To calculate the expected *outcomes, the actual at bats for both normal seasons and World Series are multiplied by this new theoretical batting average. Thus for the World Series with 98 at bats the Expected number of hits would be 98 x .263 = 25.8. Likewise, the expected number of regular season hits would be 9864 x .263 = 2594 hits during regular play. We now have two sets of outcomes to compare in a table:*

Type of Play	*Observed Hits*	*Expected Hits*
Regular Season	*2584*	*2594*
World Series	*35*	*25.8*

(Note that this is really a one-column, two-row matrix)

We calculate the χ^2 *statistic:*

$$\chi^2 = \sum \frac{(\text{Observed - Expected})^2}{\text{Expected}}$$

$$= \frac{(35-25.8)^2}{25.8} + \frac{(2584-2594)^2}{2594}$$

$$= \frac{9.2^2}{25.8} + \frac{-10^2}{2594}$$

$$= \frac{84.64}{25.8} + \frac{100}{2594}$$

$$= 3.28 + .04$$

$$= 3.32$$

To complete the chi-square test, using the chi-square table, we must determine the degrees of freedom for this value of χ^2. *For a one column (single group) set of data, degrees of freedom is the number of rows minus 1. If "r" is the number of rows (2 in our exercise), degrees of freedom = (r-1). In this exercise, degrees of freedom = (2-1) = 1. Before completing the test, we must determine the value of* α*, the level of significance, which we will apply to this test. Lets choose* $\alpha = .05$*. Now we enter the chi-square table at 1 degree of freedom (symbolized by* ν*) and go across to the column for a = .05. We find a value of* $\chi^2 = 3.84$*,* *which exceeds the value calculated above. Since the calculated χ^2 did not exceed the critical χ^2 value for this test, we accept the null hypothesis and conclude that the different batting averages between regular season and World Series do not represent true differences in performance.*

Exercise 9.3
A group of 25 students have the following ages:
25, 22, 22, 27, 20, 21, 20, 24, 24, 26, 21, 21, 20, 22, 26, 21, 22, 28, 22, 25, 25, 21, 22, 25, 24
Can this data be interpreted using the normal distribution?

We prepare a table to allow us to use the chi-square test.

Integer Value	*Observed Count*	*Normal Proportion*	*Expected Count*
20	3		
21	5		
22	6		
23	0		
24	3		
25	4		
26	2		
27	1		
28	1		

First we must calculate the mean and standard deviation for this data:

$\bar{x} = 23.04$, $s = 2.34$

The suggested restrictions that apply to the chi-square test are: (a) all expected values are equal to or greater than 2 and (b) half or more of expected values are greater than 5. In order to meet his criteria, it is necessary to combine several cells: 22 and 23; 24 and 25; and 26, 27 and 28.

Integer Value	*Observed Count*	*Normal Proportion*	*Expected Count*
20	*3*		
21	*5*		
22 to 23	*6*		
24 to 25	*7*		
26 to 28	*4*		

We will now calculate the Expected values of the cells from the normal curve with $\bar{x}$ = *23.04, s = 2.34. When we have an integer value (of the observed data) equal to 20 what do we mean? We would obtain an integer of 20 by rounding any data in the range from 19.5 to 20.49. Because the normal curve is a continuous distribution, our integer values have to be converted into the correct range of continuous data. Thus any integer value i represents a (continuous) range of values from (i - .5) to (i + .49). The first value in our data set (20) and the final value (26 to 28) will actually begin at* $-\infty$ *and extend to* $+\infty$ *since the normal curve extends to both values. The normal proportions are calculated from the Z scores for these groups of values as follows:*

For 20, normal distribution range is $-\infty$ *to 20.49, this corresponds to a value of* $Z = \frac{20.49 - 23.04}{2.34}, = \frac{-2.55}{2.34}$ *= -1.09. The proportion of the normal curve from* $-\infty$ *to –1.09 = 0.5000 - .3621 = .1379 which when multiplied by 25 (total number of students) gives an Expected count of 3.45*

For 21, normal data range is 20.5 to 21.49, this corresponds to a value of $Z = \frac{21.49 - 23.04}{2.34}, = \frac{-1.55}{2.34}$ *= -0.66. The proportion of the normal curve from –1.09 to -0.66 = .3621 - .2454 = .1167 which when multiplied by 25 gives an Expected count of 2.92.*

For 22 to 23, normal data range is 21.5 to 23.49, this corresponds to a value of $Z = \frac{23.49 - 23.04}{2.34}, = \frac{0.45}{2.34} = +0.19$. *The proportion of the normal curve from -0.66 to +0.19 = .2454 + .0753 = .3207 which when multiplied by 25 gives an Expected count of 8.02.*

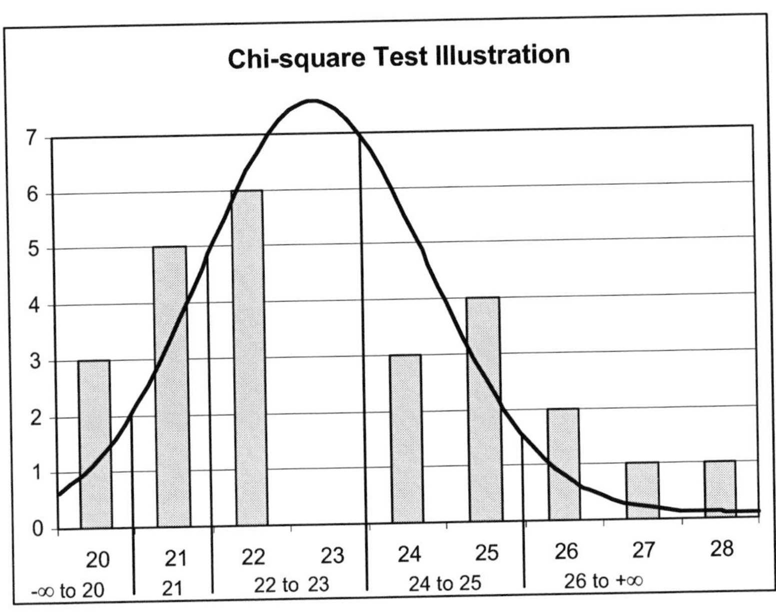

For 24 to 25, normal data range is 23.5 to 25.49, this corresponds to a value of $Z = \frac{25.49 - 23.04}{2.34}, = \frac{2.45}{2.34} = +1.05$. *The proportion of the normal curve from 0.19 to 1.05 = - .0753 + .3531 = .2778 which when multiplied by 25 gives an Expected count of 6.95.*

For 26 to 28, normal data range is 25.5 to +∞, this corresponds to a value of $Z > \frac{25.49 - 23.04}{2.34}, > \frac{2.45}{2.34} > 1.05$. *The proportion of the normal curve greater than 1.05 = -.3531 + .5000 = .1469 which when multiplied by 25 gives an Expected count of 3.67*

Integer Value	*Observed Count*	*Normal Proportion*	*Expected Count*
20	*3*	*.1379*	*3.45*
21	*5*	*.1167*	*2.92*
22 to 23	*6*	*.3207*	*8.02*
24 to 25	*7*	*.2778*	*6.95*
26 to 28	*4*	*.1469*	*3.67*

It is now possible to calculate the χ^2 statistic and complete the test.

$$\chi^2 = \sum \frac{(\text{Observed - Expected})^2}{\text{Expected}}$$

$$= \frac{(3-3.45)^2}{3.45} + \frac{(5-2.92)^2}{2.92} + \frac{(6-8.02)^2}{8.02} + \frac{(7-6.95)^2}{6.95} + \frac{(4-3.67)^2}{3.67}$$

$$= \frac{.2025}{3.95} + \frac{4.3264}{2.92} + \frac{4.0804}{8.02} + \frac{.0025}{6.95} + \frac{.1089}{3.67}$$

$$= .052 + 1.48 + 0.509 + 0.0003 + .030 = 2.0713.$$

Degrees of freedom for this one column test = (r-1) = (5-1) = 4, for α = .05 the critical χ^2 = 9.488. Since the calculated chi-square is less than the critical value, we accept the null hypothesis that the normal curve fits this data.

Note that the above example has used the non-parametric Chi-square test to confirm that further examination of these ages could be done using parametric tests.

Sign Test of Significance

The sign test of significance is a non-parametric test that requires paired data, either matched data for different treatments or matched before and after data. In the former case, each pair is matched via criteria that relate to the issue being tested. If comparing two different methods of teaching swimming, the pairs would be matched on their swimming ability before the treatment. As a non-parametric test, the sign test requires no knowledge of underlying population distributions and is solely dependent upon simple probabilities. Fundamentally the test involves evaluating the direction of change in a characteristic of the matched pairs, either positive or negative. Evaluation is based upon the probabilities of the binomial distribution for $p = 0.5$ (the same probabilities as a coin toss).

Sign Test Requirements

1. Ordinal Data
2. Two-group test
3. Related groups
4. When a pair of observations are tied, neither is used
5. Plus and minus signs are used to indicate differences
6. Primarily a one-tail test (Two-tail probabilities are obtained by doubling the table values – Appendix 5)

The test can best be explained by illustrating with an example. Suppose a commercial weight reduction program wants to show the value of their program by tracking weight changes after a 6-week treatment. For each participant the initial weight is recorded and then the final weight. The data for the 12 participants might look something like:

Participant	Initial Weight	Final Weight	Sign (loss = -)
A	160	161	+
B	230	200	-
C	142	136	-
D	155	140	-
E	220	221	+
F	136	135	-
G	172	172	0

H	300	240	-
I	148	143	-
J	179	181	+
K	193	192	-
L	140	135	-

Looking at the above table, each change is labeled as either – (a loss), 0 (no change), or + (a gain). The null hypothesis is that there is no significant difference between before and after for this sample. The implication of this null hypothesis is that all changes are random and the distribution of changes (like coin tosses) can be attributed to random chance with a median occurrence of 0.5. To perform the test, first ignore all "0" changes; second add up the minuses and pluses (# of - = 8), (# of + = 3); determine the sample size n = number of matched pairs less all zero results (n = 12 – 1) = 11; finally determine the level of significance for the test, $\alpha = 0.10$ (for this example). Either using the binomial table (Appendix 3) for p = 0.5, or the special Sign Test table (Appendix 5) determine the probability of the result obtained. The (Appendix 5) table is always entered for the lesser value obtained.

In this example the test is determining the probability of obtaining as few as 8 weight losses. (Another way to say this is what is the probability of 8 or more weight losses from a sample of 11 weight changes?) Since the weight losses are the larger number, we use the table to determine the (equivalent) probability of three or less weight gains. For a sample size of 11, the probability of 8 or more successes is the same as the probability of 3 or less non-successes. If the probability from the (Appendix 5) table is less than α reject the null hypothesis, if the probability from the table is greater than or equal to α, accept the null hypothesis.

If we used the binomial table to evaluate this example, we would have to add together the probabilities for 3 or less successes for n = 11 trials at p = 0.5. From the (Appendix 3) table we get:

1 out of 11, probability = .0054
2 out of 11, probability = .0269
3 out of 11, probability = .0806
Total probability = .1129

Or if we used the Sign Test table (Appendix 5) we would enter the column at x = 3 and go down to the row for n = 11. The number at the intersection is 0.1133, which is the same as the result above (with some rounding error). This table eliminates the need to add the various probabilities that the Binomial Table requires. When n > 20, the normal distribution (Z table) can be used as was referenced in the section on the binomial distribution.[6]

Since the probability of this result exceeds our level of significance, we accept the null hypothesis. This may be a difficult result to accept, since by inspection it sure looks like a lot of weight loss. The fact that the sign test can't discriminate this result more finely reflects the relatively low power of the sign test. Low power indicates that the test is less able to discriminate for the same amount of data as a more powerful test. We will later apply this same data to more powerful tests as an example of their greater discrimination.

Exercise 9.4
A study of two different swimming training methods was conducted using 30 students. Each student was rated by an independent judge for attitude toward water and swimming on a scale of 1 to 10. The students were then split into two groups with students of equal initial scores arbitrarily assigned to Group A and Group B. Group A was taught to swim by "total immersion" method – sink or swim. Group B was taught through water games and utilization of learning aids. After six weeks of "classes", each student was again evaluated on attitude towards water and swimming on the same ten-point scale by an independent judge. The results are shown below. Formulate the null hypothesis and perform the sign test using a level of significance, $\alpha = 0.05$.

---

[6] Sharp, Vicki F. (1979), *Statistics for the Social Sciences*. Boston, MA: Little Brown and Company (Library of Congress Catalog Card No. 78-70849), page 235.

Matched Pair ID	*Initial Score*	*Final Score -A*	*Final Score - B*	*Sign (B – A)*
1	1	2	4	+
2	1	3	4	+
3	2	2	3	+
4	2	4	4	0
5	2	3	4	+
6	3	5	6	+
7	3	2	3	+
8	3	3	4	+
9	4	5	4	-
10	5	4	6	+
11	5	3	3	0
12	5	5	6	+
13	6	4	7	+
14	7	8	6	-
15	9	8	9	+

The data necessary to perform the sign test are all present in the table. The null hypothesis statement would be H_0: there is no significant difference between the results of the two methods of teaching swimming. The values for the sign test are: x (smallest number of successes or failures) = 2, n (number of trials) = 13 (total 15 – 2 zeroes). From the table in Appendix 5, the binomial probability associated with these values = 0.0112, is less than the 0.05 level of significance. Therefore we can reject the null hypothesis and accept an alternate hypothesis that the two teaching method results are not equivalent.

The sign test table is inherently one-tailed. For a two-tailed evaluation, the value of probability obtained from the table is doubled. Thus for a two-tailed test at $\alpha = 0.05$, the value of 0.0112 obtained above is doubled to a value of 0.0224. In this circumstance, the data would also pass a two-tailed test.

Wilcoxon Signed-Ranks Test

The sign test has lower power than other tests because it ignores any additional data beyond the direction of change in the situation. It only depends upon the direction of change (sign) and ignores (where it exists) the magnitude of change. The Wilcoxon Signed-Ranks test is a more powerful test for the same types of data that does take the magnitude of change into account. Before discussing the details of the test it is important to understand how to rank order sample data.

Ranks are numbers assigned to people or things that order the respective data from lowest to highest or vice-versa. Ranks are assigned to scores after listing the scores from lowest to highest. What do we do about scores that have the same values? Suppose we have the following scores: 13, 12, 14, 17, 14, 19. To rank them the first act is to line them up from lowest to highest: 12, 13, 14, 14, 17, 19. Then we assign a rank number to the lined scores, from 1 to … (in this case 6).

Rank:	1	2	3	4	5	6
Score:	12	13	14	14	17	19

When we have two or more scores with the same value, their rank is the average of all the ranks assigned to the same score:

Rank:	1	2	3.5	3.5	5	6
Score:	12	13	14	14	17	19

Lining up all the scores before averaging the ranks for equal scores helps avoid losing the place in rank for all the other scores. For example the ranking of the following eight scores requires multiple averages: 130, 125, 95, 105, 105, 130, 105, 110. First do the simple ranking:

Rank:	1	2	3	4	5	6	7	8
Score:	95	105	105	105	110	125	130	130

Now average the ranks for the three values of 105 and the two values of 130:

Rank:	1	3	3	3	5	6	7.5	7.5
Score:	95	105	105	105	110	125	130	130

This type of ranking of scores is necessary for tests such as the Wilcoxon that will actually calculate values using the assigned ranks.

<u>Wilcoxon Signed-Ranks Test Requirements</u>

1. Ordinal data
2. Two groups
3. Related (paired or matched) groups
4. Ranked data
5. One- or two-tailed test

Exercise 9.5

As preparation for the details of the Wilcoxon, rank order the weight changes from the exercise used in the sign test. Perform the ranking on the absolute value of the change (initially) without regard to the sign of the change (exclude any 0 changes).

Rank:	*1*	*2*	*3*	*4*	*5*	*6*	*7*	*8*	*9*	*10*	*11*
Wt. Change:											

Reorder the ranks for duplicate scores:

Rank:
Wt. Change:

Begin preparing the data for the Wilcoxon by calculating the absolute value of the changes, as per the table following:

Participant	Initial Weight	Final Weight	Sign (loss = -)	Change Absolute Value
A	160	161	+	1
B	230	200	-	30
C	142	136	-	6
D	155	140	-	15
E	220	221	+	1
F	136	135	-	1
G	172	172	0	0
H	300	240	-	60
I	148	143	-	5
J	179	181	+	2
K	193	192	-	1
L	140	135	-	5

Then create the ranks:

Rank:	*1*	*2*	*3*	*4*	*5*	*6*	*7*	*8*	*9*	*10*	*11*
Wt. Change:	*1*	*1*	*1*	*1*	*2*	*5*	*5*	*6*	*15*	*30*	*60*

Average the ranks for equal scores:

Rank:	*2.5*	*2.5*	*2.5*	*2.5*	*5*	*6.5*	*6.5*	*8*	*9*	*10*	*11*
Wt. Change:	*1*	*1*	*1*	*1*	*2*	*5*	*5*	*6*	*15*	*30*	*60*

These ranked scores will be used to construct the Wilcoxon test. The basis of the Wilcoxon test is to add together all the ranks of the positive differences and separately all the ranks of the negative differences. Zero values are ignored. The smaller of the two sums is compared to the Wilcoxon table value as will be demonstrated. To perform the test complete the following table:

Sign (loss = -)	Change Absolute Value	Negative Ranks	Positive Ranks
+	1		2.5
-	30	10	
-	6	8	
-	15	9	
+	1		2.5
-	1	2.5	
0	0		
-	60	11	
-	5	6.5	
+	2		5
-	1	2.5	
-	5	6.5	
	Sum of Ranks	(-)56	(+)10

Before completing the test and entering the table we must finish structuring the test. The null hypothesis for this test is the same as was developed for the sign test, that there is no significant difference between before and after for this sample. We will perform the test at the same level of significance as the sign test, $\alpha = 0.05$, one-tailed. Take the sum of the ranks that is the smallest absolute value, enter the Wilcoxon table (Appendix 6) for the value n = the number of all the non-zero differences, in this case n = 11. Enter the row for n = 11 and go to the column corresponding to the level of significance (for our one-tailed $\alpha = 0.05$ test). The table value is 14. If our (smaller) sum of ranks is less than the table value, (10 <u>is</u> less than 14) then we can reject the null hypothesis. Thus we are able with the Wilcoxon test to reject the null hypothesis, whereas with the Sign test we were not. This demonstrates the greater power of the Wilcoxon over the Sign test.

There are two different ways that the data for the Wilcoxon test can be related. The first is when the same people are evaluated under two different conditions, e.g. before and after or two different treatments sequentially. The second is when pairs have been carefully matched on an equal basis before testing. When the sample size n is greater than 25, a normal distribution is used.[7]

Exercise 9.6
Perform the Wilcoxon Signed-Ranks Test on the data from the prior Sign Test exercise for two methods of teaching swimming. Use the same level of significance.

Matched Pair ID	*Final Score A*	*Final Score B*	*Sign (B – A)*	*Absolute Value*	*Rank*	*(+) Ranks*	*(-) Ranks*
1	2	4	+	2			
2	3	4	+	1			
3	2	3	+	1			
4	4	4	0	0			
5	3	4	+	1			
6	5	6	+	1			
7	2	3	+	1			
8	3	4	+	1			
9	5	4	-	1			
10	4	6	+	2			
11	3	3	0	0			
12	5	6	+	1			
13	4	7	+	3			
14	8	6	-	2			
15	8	9	+	1			

Prepare the summation of positive and negative ranks. Prepare the null hypothesis using the level of significance, $\alpha = 0.05$. Finally find the critical value from the Wilcoxon table and complete the test either accepting or rejecting the null hypothesis.

---

[7] Sharp, Vicki F. (1979), *Statistics for the Social Sciences*. Boston, MA: Little Brown and Company (Library of Congress Catalog Card No. 78-70849), page 240.

Rank = 1 2 3 4 5 6 7 8 9 10 11 12 13 14 15
Score =

Rank =
Score =

Σ +Ranks = *Σ -Ranks =*

Table value =

Conclusion:

In order to accomplish this test you have to get the ranking established:

Rank = 1 2 3 4 5 6 7 8 9 10 11 12 13 14 15
Score = 1 1 1 1 1 1 1 1 1 2 2 2 3

The re-ranking for the multiple values are as follows:

Rank = 5 5 5 5 5 5 5 5 5 11 11 11 13
Score = 1 1 1 1 1 1 1 1 1 2 2 2 3

Matched Pair ID	Final Score A	Final Score B	Sign (B – A)	Absolute Value	Rank	(+) Ranks	(-) Ranks
1	2	4	+	2	11	11	
2	3	4	+	1	5	5	
3	2	3	+	1	5	5	
4	4	4	0	0			
5	3	4	+	1	5	5	
6	5	6	+	1	5	5	
7	2	3	+	1	5	5	
8	3	4	+	1	5	5	
9	5	4	-	1	5		5
10	4	6	+	2	11	11	

11	*3*	*3*	*0*	*0*			
12	*5*	*6*	*+*	*1*	*5*	*5*	
13	*4*	*7*	*+*	*3*	*13*	*13*	
14	*8*	*6*	*-*	*2*	*11*		*11*
15	*8*	*9*	*+*	*1*	*5*	*5*	

Σ +Ranks = (+)75 *Σ -Ranks = (-)16*

The Wilcoxon table is entered at a value of n = 13, (15 – 2) at a one-tailed test level of significance α = 0.05, resulting in a table value = 21. Since the lesser sum of the ranks (16) is smaller than the table value, we reject the null hypothesis. Thus we can conclude (as we did with the sign test) that these two methods of teaching swimming are not equivalent.

Student's "t" Test of Differences Between Means

When the standard deviation of a normal population is known, it is easy to calculate confidence intervals and perform tests of significance using the Z table. When the standard deviation of the normal population is not known, it must be estimated using the sample standard deviation. Since the sample standard deviation, s, is an unbiased estimator of the population standard deviation, σ, this is a correct approach to take. However, any given value of s developed from a sample is subject to sampling error and may over or underestimate the true value of σ. If the sample standard deviation overestimates σ, the resultant confidence interval or test of significance will be overly conservative but will still be correct. If the sample standard deviation underestimates σ, then both the confidence interval and tests of significance will be incorrect. The values will be too small (based on the underestimating value of s) and the results produced will not truly be at the probability levels desired. For instance a 95% confidence interval developed from an s that underestimates σ will in reality only be an 85% or 90% interval. The potential for this kind of error is largest when sample sizes are smallest.

In 1908 a statistician working at the Guinness Brewing Company, William Gossett, developed an approach that corrects for this issue with

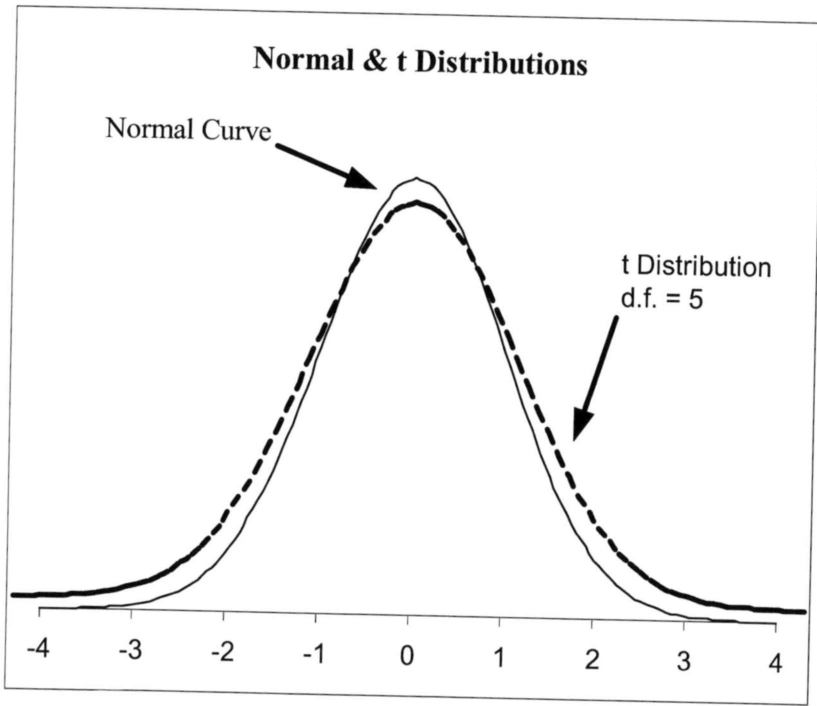

small samples. He could not publish the discovery under his own name, so he published under the pseudonym "Student". His Student's t distribution resolves this problem. When the population standard deviation is being estimated from a sample standard deviation, the relationship of probability and distance from the distribution mean is described by the "t" distribution instead of the Z (or normal) distribution. The t distribution is like the normal distribution except that it is flatter and wider. Also there is a separate t distribution for each different sample size (Appendix 7). The illustration above compares the normal and t distributions where the t distribution is for a sample with 5 degrees of freedom. (Degrees of freedom are usually the sample size minus one, df = n – 1.) As the sample size increases, the t distribution gets narrower and taller. With a sample size producing 30 degrees of freedom, the difference between the t value and the normal curve becomes quite small, and the normal curve or Z table can be applied to larger samples.

It is a requirement that the population distribution being sampled is normal to be able to use the t statistic with the sample results. However, extensive studies have shown that the t statistic is quite robust to non-normal distributions. This robustness exceeds the capability of other tests to discern non-normal populations. Therefore as long as the data is interval or ratio and the population distributions are not exceedingly non-normal, the t statistic will likely be reliable.

t-Test Requirements

1. Interval or ratio data
2. Random sample drawn from a normally distributed population
3. If two groups, they are independent
4. If two populations, they have the same variances
5. One-tailed or two-tailed tests

Single Sample Applications – Confidence Intervals

If we use a small (n < 30) sample to infer a population mean from a sample mean, we use the t table to establish confidence intervals. For example suppose that a population of students is known to be normally distributed as to age. A sample of twelve students is randomly selected and their ages recorded. The ages are: 27, 52, 43, 47, 36, 44, 33, 40, 43, 55, 47, 39. These ages are used to calculate a sample average and standard deviation as follows:

$$\bar{X} = \frac{\sum X_i}{n} = \frac{506}{12} = 42.2 \qquad s = \sqrt{\frac{\sum(\bar{X} - X_i)^2}{n-1}} = 7.86$$

In order to calculate a 95% confidence interval for the population average age, μ, we will calculate the standard deviation of the sample mean as follows:

$s_{\bar{X}} = \frac{\sigma}{\sqrt{n}}$, but since we don't know σ we substitute

$$= \frac{s}{\sqrt{n}} = \frac{7.86}{\sqrt{12}} = \frac{7.86}{3.46} = 2.27$$

If we were working with σ, the 95% confidence interval would simply be

$\bar{X} \pm 2s_{\bar{X}}$, where the 2 is the (approximate) value of Z corresponding to 95% of the normal curve. Since our value of $s_{\bar{X}}$ is calculated only from sample data, instead of Z = 2, the appropriate T value must be used. To determine this value we go to the t table for n – 1 (12 – 1 = 11) degrees of freedom and a two-tailed probability of 0.05. (This is the value of α = .05, where our confidence interval value 95% = .95 = 1 – α.) Looking at the t table we find a t value of 2.201. Thus our confidence interval ($\pm 2.201\ s_{\bar{X}}$) will be a little wider than that we would use if we knew the true value of σ ($\pm 2\ s_{\bar{X}}$). We calculate the result:

$$95\%\text{C.I.} = \bar{X} \pm 2.201\ s_{\bar{X}} = 42.2 \pm 2.201(2.27) = 42.2 \pm 5.0$$

$$= 37.2 \text{ to } 47.2$$

Single Sample Applications – Test of the Null Hypothesis

Suppose we know that the average age of adult college students is 36.4 based on census data, however, the population standard deviation is not known. Is the sample of students we developed above representative of this population? The t-test is used to compare the differences between means when the sample size is less than 30. Begin by forming the null hypothesis, H_0: there is no difference between the population mean (36.4) and the sample mean (42.2). The alternate hypothesis is that this sample represents an older population that is, H_1: the sample mean (42.2) is greater than the population mean (36.4). Set the level of significance for this test, α = .05 (one tailed test). Now determine the test t value:

$t = \dfrac{\bar{x} - \mu}{s_{\bar{x}}} = \dfrac{42.2 - 36.4}{2.27} = \dfrac{5.8}{2.27} = 2.55$. If this t statistic exceeds the table t value, we reject the null hypothesis. To find the table t value, calculate degrees of freedom df = n – 1, = 12 – 1 = 11. Go to the t table for 11 degrees of freedom and a one-tailed α = .05. The value in the table is 1.796. Since our calculated value exceeds the table value, we reject the null hypothesis and accept the research hypothesis that our average student age is greater than the population age.

Exercise 9.7
In the state of Washington during the year 1997, per capita personal

income was \$26,718 with an unknown standard deviation. (It can be assumed that income is normally distributed.) A sample of 17 adult graduate students from the state was surveyed as to 1997 income. An average income of \$19,600 was reported with a standard deviation of \$9750.
a. Calculate the 90% confidence interval for student income for 1997.
b. Perform a two-tailed statistical test at the 0.05 level of significance of the null hypothesis that the student income does not differ from per capita income.

a. *Begin by calculating the standard deviation of the mean:*

$$s_{\bar{X}} = \frac{s}{\sqrt{n}} = \frac{9{,}750}{\sqrt{17}} = \frac{9{,}750}{4.123} = \$2365$$

Find the t value for df = n – 1, = 17 – 1, = 16; for a two-sided test at α = 0.1, from the table, t = 1.746,

$$90\%\ C.I. = \bar{X} \pm 1.746\ s_{\bar{X}} = \$19{,}600 \pm 1.746\ (\$2365)$$

$$= \$19{,}600 \pm \$4129 = \$15{,}471 \text{ to } \$23{,}729$$

b. *Calculate the t statistic:*

$$t = \frac{\bar{x} - \mu}{s_{\bar{x}}} = \frac{\$19{,}600 - \$26718}{\$2{,}365} = \frac{-\$7{,}118}{\$2{,}365} = -3.01$$

compare to the table value for df = n – 1 = 17 – 1 = 16, and a = .05 for a two-tailed test = 2.120. Since the (absolute) calculated value exceeds the table value, reject the null hypothesis and conclude that the student income is different than the per capita income.

Two Sample Applications – Confidence Intervals and Test of the Null Hypothesis

When the averages for two samples are being compared, and the true population standard deviation is unknown, the standard deviations of the two samples are combined in a process known as pooling to create a weighted average. This pooling process in which each value is multiplied by its associated degrees of freedom gives more priority to the larger sample, which is appropriate, since the larger sample is likely to carry more information on the population standard deviation. Since the pooling is an additive process, it is actually performed on the sample variances (the squared standard deviations) to produce a pooled variance. This avoids the equivalent issue of negative differences canceling positive differences in the combining (just as standard deviations are calculated using squared differences). The weighted variances are combined and divided by the total degrees of freedom to get the "average" variance.

$$s_p^2 = \frac{(n_1 - 1)s_1^2 + (n_2 - 1)s_2^2}{(n_1 + n_2 - 2)}$$

The square root of this pooled variance (symbolized as s_p) is the best estimate of the population standard deviation. It is used to calculate a standard deviation of the means of the two samples as follows:

$$s_{\bar{x}} = s_p\sqrt{\frac{1}{n_1} + \frac{1}{n_2}}$$

(this equation combines the standard deviations of the two means.)

This value is then used to calculate the subsequent confidence intervals and to perform tests of significance. For example if a sample is drawn from each of two different populations which have the same standard deviations, the values of the difference between the sample means can be used to calculate a confidence interval for the value of the difference between the population means.

The average weight of 15 college level football players was 237 pounds with a standard deviation of 15 pounds. Twelve NFL football players weighed an average of 253 pounds with a standard deviation 18 pounds. Create a 90% confidence interval of the average weight difference between the two populations. Assume the two populations are normally distributed with equal standard deviations.

First we must pool the two standard deviations for the samples to develop a pooled estimate of the population(s) standard deviation.

$$s_p^2 = \frac{(n_1 - 1)s_1^2 + (n_2 - 1)s_2^2}{(n_1 + n_2 - 2)} =$$

$$\frac{(15-1)15^2 + (12-1)18^2}{(15+12-2)} = \frac{(14)225 + (11)324}{(25)} = \frac{6714}{25} = 268.56$$

$$s_p = 16.39$$

Next we generate the confidence interval, but with two samples the degrees of freedom is: df = $n_1 + n_2$ - 2. This is 12 + 15 – 2 = 25, the same as the denominator in the pooled standard deviation calculation. From the table (Appendix 7), the t statistic for 25 df and α = .10 (two-tailed) is 1.708. We now construct the confidence interval on the difference between the two averages.

$$90\%\ \text{C.I.} = (\bar{X}_1 - \bar{X}_2) \pm t_{table}\ s_p\sqrt{\frac{1}{n_1} + \frac{1}{n_2}}$$

$$= (237\text{-}253) \pm 1.708\ (16.39)\sqrt{\frac{1}{12} + \frac{1}{15}}$$

$$= (237\text{-}253)\ \pm 1.708\ (16.39)\sqrt{.1500}$$

$$= (-16)\ \pm 1.708(16.39)(.387)$$

$$= (-16)\ \pm 10.8.$$

This produces a range of weight difference (90% C.I.) from –5.2 to -26.8 pounds.

The same pooling and calculation of the pooled standard deviation of the means is used for tests of hypotheses. The Exercise below illustrates this use.

Exercise 9.8
The weights of 15 New York models have a sample mean of 117 pounds with a standard deviation 10 pounds. Sixteen Los Angeles models have a mean weight of 123 pounds with a standard deviation of 8 pounds. Develop a null hypothesis that both samples come from the same overall population of models and test the hypothesis at an $\alpha = .05$ *level of significance (two-sided test).*

In order to calculate the test we must begin by constructing the null hypothesis that these apparent weight differences are not really significant. H_0: $\bar{x}_1 - \bar{x}_2 = 0$. *We also construct the alternative hypothesis* $H1$: $\bar{x}_1 \neq \bar{x}_2$. *To calculate the t statistic:*

$$t = \frac{\bar{x}_1 - \bar{x}_2}{s_{\bar{x}}}$$

we must first pool the two standard deviations for the samples to develop a pooled estimate of the population(s) standard deviation.

$$s_p^2 = \frac{(n_1 - 1)s_1^2 + (n_2 - 1)s_2^2}{(n_1 + n_2 - 2)} =$$

$$\frac{(15-1)10^2 + (16-1)8^2}{(15+16-2)} = \frac{(14)100 + (15)64}{(29)} = \frac{2360}{29} = 81.38$$

$s_p = 9.02$

Next we generate the standard deviation of the combined means using:

$$s_{\bar{X}} = s_p\sqrt{\frac{1}{n_1}+\frac{1}{n_2}} = 9.02\sqrt{\frac{1}{15}+\frac{1}{16}} = 9.02\sqrt{.1292}$$

$$= 9.02(.359) = 3.24$$

Now we can calculate the t statistic:

$$t = \frac{\bar{X}_1 - \bar{X}_2}{s_{\bar{X}}} = \frac{117-123}{3.24} = -1.85$$

(the negative sign does not affect the t value)

Complete the test by comparing to the table value for df = $n_1 + n_2 - 2$ = 15 + 16 – 2 = 29 and a level of confidence a = .05 (two-sided). The t value from the table is 2.045. Since the calculated value does not exceed the table value, we accept the null hypothesis that the two samples are not significantly different.

F Test of Sample Variance

We begin this section with a caution. The F test applied to the variance of two samples as a test of the null hypothesis is extremely sensitive to any non-normality in underlying populations. This behavior is described as non-robust. Even slight departures from normality can result in test performance that is dramatically incorrect. An example would be a test performed at an $\alpha = 0.05$ level producing a result with actual probabilities ranging from $\alpha = 0.01$ to $\alpha = 0.10$.[8] These results have led another text author to advise regarding the F test for variances: "Our advice here is short and clear: don't do it". Given this background

[8] E. S. Pearson and N. W. Please, "Relation between the shape of population distribution and the robustness of four simple test statistics," *Biometrika*, **62**, (1975), pp 223-241.

you may well ask, "Why present this material?" The answer lies in the other application of the F tables, the Analysis of Variance (ANOVA). This procedure, the final presentation in this text, while it relies on the same tables of values is quite robust and effective in comparing the means of three or more samples.

Assume you have a circumstance in which two samples have been prepared in a manner that could affect their variability. While the standard deviations for the samples do indeed differ, the question is do they differ enough that they represent truly different levels of variation? The F test will provide an answer to this question. Again it does so at the risk that any non-normality in the underlying populations of either sample will affect the validity of the outcome. Nonetheless, this test can be used to answer the above question. It is a simple test involving creating a ratio of the variances (squared standard deviations) for the two outcomes and comparing to critical values for the related degrees of freedom of both samples. Because the result is reflective of the separate degrees of freedom of the numerator and the denominator, the F table is quite complex and extensive. An example will demonstrate its use.

<u>F-Test Requirements</u>

1. Interval or ratio data
2. Independent samples from normally distributed populations
3. One-tailed test

Two different forms of IQ test are being developed. The primary concern is whether they display the same variability in results when applied to the same subjects. Form A is applied to 20 people and the sample standard deviation is 11 "IQ units". The second (Form B) is applied to 22 people and the sample standard deviation is 17 "IQ units". Assuming that these results are both from normally distributed populations, test H_0 at the 0.05 level (two-sided test).

Begin by constructing H_0: $s_1 = s_2$, and H_1: $s_1 \neq s_2$. To prepare to perform the test with $\alpha = .05$ (two-tailed), go to the table for numerator df = 22 – 1 = 21, denominator df = 20 – 1 = 19 (the larger variance value is always used for the numerator). Because the F test is inherently one-sided, cutting the confidence interval in half approximates the two-sided equivalent. Thus we go to the table for df = 21 for the numerator and df =

19 for the denominator at an $\alpha = 0.025$. The critical value of the F ratio is 2.49. Now calculate the F ratio for the data:

$$F = \frac{s^2_{larger}}{s^2_{smaller}} = \frac{(17)^2}{(11)^2} = \frac{289}{121} = 2.39$$

Since the calculated F ratio does not exceed the critical table value, we accept the null hypothesis that these standard deviations really are equivalent.

Exercise 9.9

An experimenter is studying the capabilities of novice and experienced female rowers. A key characteristic is how rapidly the knee straightens as the rower pushes back with her legs. The study produced the following variability information:

Group	*n*	*s*
Novices	*8*	*.96*
Skilled	*10*	*.48*

Assuming these samples are from normal populations, are the standard deviations equivalent at an $\alpha = .05$, one-sided level of confidence?

We will be comparing the F ratio we calculate with the table value for df numerator = 7 and df denominator = 9 at an $\alpha = 0.05$ (one-sided). The operative hypotheses are: H_0: $s_{novice} = s_{skilled}$, and H_1: $s_{novice} > s_{skilled}$. The critical F ratio value from the table is 3.29. Calculate the F ratio for the two standard deviations:

$$F = \frac{s^2_{larger}}{s^2_{smaller}} = \frac{(.96)^2}{(.48)^2} = \frac{.9216}{.2304} = 4.0$$

because this exceeds the table F value, we reject the null hypothesis and conclude that the novices show a larger variation than the skilled rowers.

Analysis of Variance (ANOVA)

The t statistic allows us to compare the means of two samples and test the null hypothesis to determine if the means represent the same population distribution. Sometimes we want to compare the means of three or more samples and determine if we can accept the null hypothesis that all means are essentially equal. Such a comparison is accomplished using the one-way analysis of variance, often-abbreviated ANOVA.

Conceptually, the analysis of variance is pretty easy to understand. The variance within the groups (around their respective means) is compared to the variance across the groups. The two different variances will be compared to develop an F ratio. That F ratio is compared to the critical F ratio from the F table for whatever level of significance is pre-selected. The original measurement consists of three or more "groups" of data. The groups are the divisions of the data that we wish to test for equality of the means.

Analysis of Variance Requirements

1. Interval or ratio data
2. Normally distributed populations
3. Equal variances
4. Three or more independent groups
5. Random samples
6. One-tailed test

For example an instructor is experimenting with three different methods to teach statistics: audio tapes with no support written materials, video tapes with supporting text, and live lecture with opportunity to ask questions. An initial class of thirty students is divided randomly across the three teaching methods abbreviated "audio", "video" and "live" and all surviving students take a standard test at the end of the semester. The data for the test results is as follows (perfect score is 38):

Audio	Video	Live
38	38	38
29	33	31
26	22	29
33	28	34
31	36	35
22	31	27
25	28	36
*	31	32
*	34	36
*	*	37

*-Student did not "survive"

$\bar{X}$ =	29.1	31.2	33.5

Grand Average = 31.5

s =	5.40	4.82	3.63
n =	7	9	10

The first variance to be developed is that within the groups. This is accomplished by pooling the data for all three groups. We did simple pooling when we combined the variances for two groups to perform the t test previously.

$$s_p^2 = \frac{(n_A - 1)s_A^2 + (n_V - 1)s_V^2 + (n_L - 1)s_L^2}{(n_A + n_V + n_L - 3)}$$

$$= \frac{(7-1)5.40^2 + (9-1)4.82^2 + (10-1)3.63^2}{(7+9+10-3)}$$

$$= \frac{(6)(29.16) + (8)(23.23) + (9)(13.18)}{(23)} = \frac{174.96 + 185.84 + 118.62}{(23)}$$

$$= \frac{479.42}{23} = 20.84$$

(this is equivalent to a pooled standard deviation of 4.57)

The second variance to be developed is that between groups. This is calculated by adding the squared difference of each average to the grand average, weighted by the sample size. The degrees of freedom used in this calculation is the number of groups minus one.

$$s_b^2 = \frac{n_A(\bar{x}_A - \bar{x}_{GA})^2 + n_V(\bar{x}_V - \bar{x}_{GA})^2 + n_L(\bar{x}_L - \bar{x}_{GA})^2}{d.f.}$$

$$= \frac{7(29.1 - 31.5)^2 + 9(31.2 - 31.5)^2 + 10(33.5 - 31.5)^2}{(3-1)}$$

$$= \frac{7(-2.4)^2 + 9(-0.3)^2 + 10(2.0)^2}{2}$$

$$= \frac{7(5.76) + 9(.09) + 10(4.00)}{2} = \frac{40.32 + .81 + 40.00}{2} = \frac{81.13}{2} = 40.57$$

The F ratio is calculated by dividing the between-group variance by the within-(pooled)-group variance:

$$F = \frac{s^2_{\text{between groups}}}{s^2_{\text{within/pooled}}} = \frac{40.57}{20.84} = 1.95$$

To apply this F ratio to this data, we begin by developing the null hypothesis, H_0: $\bar{x}_A = \bar{x}_V = \bar{x}_L$ and the associated alternate hypothesis, H_1: all three averages are not equal. Note that the outcome of this test will either allow us to assume that all three averages are equal, or that all three are not. The results will not tell us which averages are not equal, only that all three are not equal. The value of the F ratio calculated is compared to the F table critical value for the df between groups (3 – 1) = 2 and the df within groups (7 + 9 + 10 – 3) = 23 for our respective level of significance. In this case let's use $\alpha = .05$, this yields a table F value of

3.42. Since the calculated F ratio does not exceed the table value, we accept the null hypothesis.

The analysis of variance requires interval or ratio data and the assumption that the populations from which the samples are drawn are normally distributed. It is, however, a very robust test – like the t test. Even though the actual test statistic is the F ratio, the one-way ANOVA is much more robust than the test of variances for equality we previously studied.

Exercise 8.14
A television repair company wanted to test the effectiveness of its three new television repair schools. Five pupils were randomly selected from each school's six-month training program. The student's results on a standardized test are shown:

	Group A	*Group B*	*Group C*
	2	2	6
	1	2	5
	2	3	5
	3	2	5
	2	3	6
$\bar{x}$	2.0	2.4	5.4
Grand Average = 3.27			
s	0.71	0.55	0.55

Perform a one-way analysis of variance for this data. Create and test the null hypothesis at an $\alpha = 0.01$ level of significance.

The null hypothesis for this situation is that all three averages are equivalent, H_0: $\bar{x}_A = \bar{x}_B = \bar{x}_C$. The alternate hypotheses, H_1 is that all three are not equal.

The first variance to be developed is that within the groups. This is accomplished by pooling the data for all three groups. (Since the groups are equal in size we do not really need to perform the pooling, however, in the interest of practicing a discipline that fits all circumstances, we will pool.)

$$s_p^2 = \frac{(n_A - 1)s_A^2 + (n_B - 1)s_B^2 + (n_C - 1)s_C^2}{(n_A + n_B + n_C - 3)}$$

$$= \frac{(5-1)0.71^2 + (5-1)0.55^2 + (5-1)0.55^2}{(5+5+5-3)}$$

$$= \frac{(4)(.5041) + (4)(.3025) + (4)(.3025)}{(12)} = \frac{2.5205 + 1.21 + 1.21}{(12)}$$

$$= \frac{4.9405}{12} = 0.412$$

(this is equivalent to a pooled standard deviation of 0.64)

The second variance to be developed is that between groups. This is calculated by calculating the squared difference of each average to the grand average, weighted by the sample size (this time the weighting is necessary).

$$s_b^2 = \frac{n_A(\bar{x}_A - \bar{x}_{GA})^2 + n_B(\bar{x}_B - \bar{x}_{GA})^2 + n_C(\bar{x}_C - \bar{x}_{GA})^2}{\text{d.f.}}$$

$$= \frac{5(2.0 - 3.27)^2 + 5(2.4 - 3.27)^2 + 5(5.4 - 3.27)^2}{(3-1)}$$

$$= \frac{5(-1.27)^2 + 5(-0.87)^2 + 5(2.13)^2}{2}$$

$$= \frac{5(1.61) + 5(.76) + 5(4.54)}{2} = \frac{8.05 + 3.80 + 22.7}{2} = \frac{34.55}{2} = 17.27$$

The F ratio is calculated by dividing the between group variance by the within (pooled) group variance:

$$F = \frac{s^2_{\text{between groups}}}{s^2_{\text{within/pooled}}} = \frac{17.27}{0.412} = 41.9$$

The table critical F value for 2 df in the numerator and 12 df in the denominator at a = 0.01 is 6.93. Since our calculated F ratio is (far) larger than the critical value, we reject the null hypothesis and accept the alternate hypothesis. Please note: which of the means: A, B or C is unequal is not indicated by this test, only that all three are not equivalent.

Appendix 1
Table of Five Digit Random Numbers

92947	91202	20234	41332	43791	92506	21829	99007	66491	29740
04583	30797	74779	21981	94336	94103	04658	62916	20303	21653
74246	93292	38258	49549	44530	48752	15184	45149	28633	29674
49723	51280	37294	83313	40891	9311	65209	43244	22253	66138
66135	29926	26514	14565	67959	44863	18334	78566	19407	18478
61173	24330	24122	53355	49550	06927	38386	18096	25476	03972
50126	34195	55403	98222	42906	02914	81171	38312	09614	21351
09744	60084	38259	17159	24742	65391	28769	41208	50669	10908
20924	83698	74703	01920	72486	12682	70593	74509	55285	21519
17982	84777	23436	03452	95314	04590	80011	27936	86244	39056
38908	81562	84753	96015	02064	83309	72219	62575	51231	95889
97197	22327	62305	39348	97212	11911	33763	23807	88810	32224
58478	37200	80920	11369	03318	99581	98623	43038	49132	77612
78450	11033	45544	54366	24940	89421	90711	90324	58280	93915
78869	12489	59616	58582	23942	37675	82315	11288	06363	84247
73421	07978	11759	60186	21224	28154	67825	55223	15338	64545
19341	62305	47728	16539	78179	57250	26569	78933	46152	26924
33510	18034	73261	71719	40032	70361	82400	99293	33365	35609
77234	52780	87438	30004	91621	47955	21365	5427	80163	94793
87331	66823	76952	73970	69306	92124	80355	40216	10551	21728
78475	01289	37937	28706	02170	70643	32731	17352	22278	08014
27307	35995	86154	09140	10924	49953	65414	90922	69603	49521
60964	36589	10011	50931	30010	21913	34711	97385	71019	92155
85803	07267	84654	34311	49171	81512	13762	54770	56688	71860
94792	02797	58505	44938	42605	34286	41586	62908	11968	29212
15508	13644	01852	24446	20742	03462	06627	22013	85521	30502
94258	78523	71669	25400	38449	37635	20580	04276	76111	90831
47171	67521	16867	69035	59866	79650	59470	19432	82425	10871
37144	57582	98181	25749	34875	78018	29239	26923	78750	02185
29252	79952	90221	88420	01341	54860	45976	06845	61622	46527
99941	35982	23105	81285	34972	01936	10129	49505	84689	28771
96284	77821	98478	50451	92916	43598	59253	76727	47984	53775
84241	32060	05754	84053	61894	87154	34229	00969	80224	53585
08872	12472	02025	04336	45992	89892	11045	17126	49394	68010
35246	30121	42491	98974	28111	49456	01591	34394	33725	44538

Appendix 2
Areas of the Standard Normal Distribution

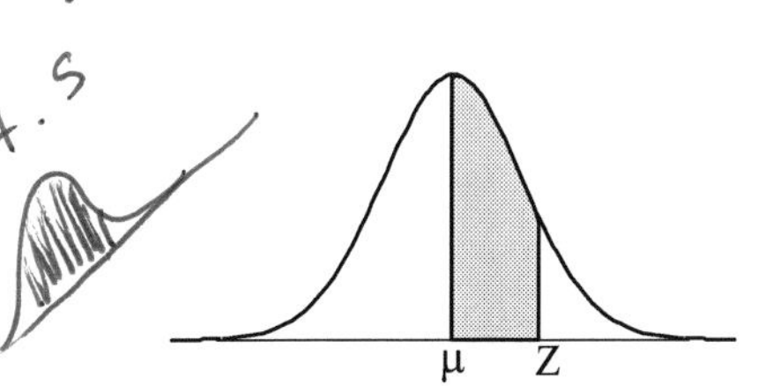

Values correspond to the area from μ to Z

Where $Z = \dfrac{X - \mu}{\sigma}$

Z	.00	.01	.02	.03	.04	.05	.06	.07	.08	.09
0.0	0.0000	0.0040	0.0080	0.0120	0.0160	0.0199	0.0239	0.0279	0.0319	0.0359
0.1	0.0398	0.0438	0.0478	0.0517	0.0557	0.0596	0.0636	0.0675	0.0714	0.0753
0.2	0.0793	0.0832	0.0871	0.0910	0.0948	0.0987	0.1026	0.1064	0.1103	0.1141
0.3	0.1179	0.1217	0.1255	0.1293	0.1331	0.1368	0.1406	0.1443	0.1480	0.1517
0.4	0.1554	0.1591	0.1628	0.1664	0.1700	0.1736	0.1772	0.1808	0.1844	0.1879
0.5	0.1915	0.1950	0.1985	0.2019	0.2054	0.2088	0.2123	0.2157	0.2190	0.2224
0.6	0.2257	0.2291	0.2324	0.2357	0.2389	0.2422	0.2454	0.2486	0.2517	0.2549
0.7	0.2580	0.2611	0.2642	0.2673	0.2704	0.2734	0.2764	0.2794	0.2823	0.2852
0.8	0.2881	0.2910	0.2939	0.2967	0.2995	0.3023	0.3051	0.3078	0.3106	0.3133
0.9	0.3159	0.3186	0.3212	0.3238	0.3264	0.3289	0.3315	0.3340	0.3365	0.3389
1.0	0.3413	0.3438	0.3461	0.3485	0.3508	0.3531	0.3554	0.3577	0.3599	0.3621
1.1	0.3643	0.3665	0.3686	0.3708	0.3729	0.3749	0.3770	0.3790	0.3810	0.3830
1.2	0.3849	0.3869	0.3888	0.3907	0.3925	0.3944	0.3962	0.3980	0.3997	0.4015
1.3	0.4032	0.4049	0.4066	0.4082	0.4099	0.4115	0.4131	0.4147	0.4162	0.4177
1.4	0.4192	0.4207	0.4222	0.4236	0.4251	0.4265	0.4279	0.4292	0.4306	0.4319
1.5	0.4332	0.4345	0.4357	0.4370	0.4382	0.4394	0.4406	0.4418	0.4429	0.4441
1.6	0.4452	0.4463	0.4474	0.4484	0.4495	0.4505	0.4515	0.4525	0.4535	0.4545
1.7	0.4554	0.4564	0.4573	0.4582	0.4591	0.4599	0.4608	0.4616	0.4625	0.4633
1.8	0.4641	0.4649	0.4656	0.4664	0.4671	0.4678	0.4686	0.4693	0.4699	0.4706
1.9	0.4713	0.4719	0.4726	0.4732	0.4738	0.4744	0.4750	0.4756	0.4761	0.4767
2.0	0.4772	0.4778	0.4783	0.4788	0.4793	0.4798	0.4803	0.4808	0.4812	0.4817
2.1	0.4821	0.4826	0.4830	0.4834	0.4838	0.4842	0.4846	0.4850	0.4854	0.4857
2.2	0.4861	0.4864	0.4868	0.4871	0.4875	0.4878	0.4881	0.4884	0.4887	0.4890
2.3	0.4893	0.4896	0.4898	0.4901	0.4904	0.4906	0.4909	0.4911	0.4913	0.4916
2.4	0.4918	0.4920	0.4922	0.4925	0.4927	0.4929	0.4931	0.4932	0.4934	0.4936
2.5	0.4938	0.4940	0.4941	0.4943	0.4945	0.4946	0.4948	0.4949	0.4951	0.4952
2.6	0.4953	0.4955	0.4956	0.4957	0.4959	0.4960	0.4961	0.4962	0.4963	0.4964
2.7	0.4965	0.4966	0.4967	0.4968	0.4969	0.4970	0.4971	0.4972	0.4973	0.4974
2.8	0.4974	0.4975	0.4976	0.4977	0.4977	0.4978	0.4979	0.4979	0.4980	0.4981
2.9	0.4981	0.4982	0.4982	0.4983	0.4984	0.4984	0.4985	0.4985	0.4986	0.4986
3.0	0.4987	0.4987	0.4987	0.4988	0.4988	0.4989	0.4989	0.4989	0.4990	0.4990
3.1	0.4990	0.4991	0.4991	0.4991	0.4992	0.4992	0.4992	0.4992	0.4993	0.4993
3.2	0.4993	0.4993	0.4994	0.4994	0.4994	0.4994	0.4994	0.4995	0.4995	0.4995
3.3	0.4995	0.4995	0.4995	0.4996	0.4996	0.4996	0.4996	0.4996	0.4996	0.4997
3.4	0.4997	0.4997	0.4997	0.4997	0.4997	0.4997	0.4997	0.4997	0.4997	0.4998
3.5	0.4998									
4.0	0.499968									

Appendix 3
Table of Binomial Probabilities

		p								
n	# of successes	0.01	0.02	0.03	0.04	0.05	0.06	0.07	0.08	0.09
2	0	0.9801	0.9604	0.9409	0.9216	0.9025	0.8836	0.8649	0.8464	0.8281
	1	0.0198	0.0392	0.0582	0.0768	0.0950	0.1128	0.1302	0.1472	0.1638
	2	0.0001	0.0004	0.0009	0.0016	0.0025	0.0036	0.0049	0.0064	0.0081
3	0	0.9703	0.9412	0.9127	0.8847	0.8574	0.8306	0.8044	0.7787	0.7536
	1	0.0294	0.0576	0.0847	0.1106	0.1354	0.1590	0.1816	0.2031	0.2236
	2	0.0003	0.0012	0.0026	0.0046	0.0071	0.0102	0.0137	0.0177	0.0221
	3	0.0000	0.0000	0.0000	0.0001	0.0001	0.0002	0.0003	0.0005	0.0007
4	0	0.9606	0.9224	0.8853	0.8493	0.8145	0.7807	0.7481	0.7164	0.6857
	1	0.0388	0.0753	0.1095	0.1416	0.1715	0.1993	0.2252	0.2492	0.2713
	2	0.0006	0.0023	0.0051	0.0088	0.0135	0.0191	0.0254	0.0325	0.0402
	3	0.0000	0.0000	0.0001	0.0002	0.0005	0.0008	0.0013	0.0019	0.0027
	4	0.0000	0.0000	0.0000	0.0000	0.0000	0.0000	0.0000	0.0000	0.0001
5	0	0.9510	0.9039	0.8587	0.8154	0.7738	0.7339	0.6957	0.6591	0.6240
	1	0.0480	0.0922	0.1328	0.1699	0.2036	0.2342	0.2618	0.2866	0.3086
	2	0.0010	0.0038	0.0082	0.0142	0.0214	0.0299	0.0394	0.0498	0.0610
	3	0.0000	0.0001	0.0003	0.0006	0.0011	0.0019	0.0030	0.0043	0.0060
	4	0.0000	0.0000	0.0000	0.0000	0.0000	0.0001	0.0001	0.0002	0.0003
	5	0.0000	0.0000	0.0000	0.0000	0.0000	0.0000	0.0000	0.0000	0.0000
6	0	0.9415	0.8858	0.8330	0.7828	0.7351	0.6899	0.6470	0.6064	0.5679
	1	0.0571	0.1085	0.1546	0.1957	0.2321	0.2642	0.2922	0.3164	0.3370
	2	0.0014	0.0055	0.0120	0.0204	0.0305	0.0422	0.0550	0.0688	0.0833
	3	0.0000	0.0002	0.0005	0.0011	0.0021	0.0036	0.0055	0.0080	0.0110
	4	0.0000	0.0000	0.0000	0.0000	0.0001	0.0002	0.0003	0.0005	0.0008
	5	0.0000	0.0000	0.0000	0.0000	0.0000	0.0000	0.0000	0.0000	0.0000
	6	0.0000	0.0000	0.0000	0.0000	0.0000	0.0000	0.0000	0.0000	0.0000
7	0	0.9321	0.8681	0.8080	0.7514	0.6983	0.6485	0.6017	0.5578	0.5168
	1	0.0659	0.1240	0.1749	0.2192	0.2573	0.2897	0.3170	0.3396	0.3578
	2	0.0020	0.0076	0.0162	0.0274	0.0406	0.0555	0.0716	0.0886	0.1061
	3	0.0000	0.0003	0.0008	0.0019	0.0036	0.0059	0.0090	0.0128	0.0175
	4	0.0000	0.0000	0.0000	0.0001	0.0002	0.0004	0.0007	0.0011	0.0017
	5	0.0000	0.0000	0.0000	0.0000	0.0000	0.0000	0.0000	0.0001	0.0001
	6	0.0000	0.0000	0.0000	0.0000	0.0000	0.0000	0.0000	0.0000	0.0000
	7	0.0000	0.0000	0.0000	0.0000	0.0000	0.0000	0.0000	0.0000	0.0000
8	0	0.9227	0.8508	0.7837	0.7214	0.6634	0.6096	0.5596	0.5132	0.4703
	1	0.0746	0.1389	0.1939	0.2405	0.2793	0.3113	0.3370	0.3570	0.3721
	2	0.0026	0.0099	0.0210	0.0351	0.0515	0.0695	0.0888	0.1087	0.1288
	3	0.0001	0.0004	0.0013	0.0029	0.0054	0.0089	0.0134	0.0189	0.0255
	4	0.0000	0.0000	0.0001	0.0002	0.0004	0.0007	0.0013	0.0021	0.0031
	5	0.0000	0.0000	0.0000	0.0000	0.0000	0.0000	0.0001	0.0001	0.0002
	6	0.0000	0.0000	0.0000	0.0000	0.0000	0.0000	0.0000	0.0000	0.0000
	7	0.0000	0.0000	0.0000	0.0000	0.0000	0.0000	0.0000	0.0000	0.0000
	8	0.0000	0.0000	0.0000	0.0000	0.0000	0.0000	0.0000	0.0000	0.0000

Appendix 3
Table of Binomial Probabilities (Continued)

		p								
n	# of successes	0.01	0.02	0.03	0.04	0.05	0.06	0.07	0.08	0.09
9	0	0.9135	0.8337	0.7602	0.6925	0.6302	0.5730	0.5204	0.4722	0.4279
	1	0.0830	0.1531	0.2116	0.2597	0.2985	0.3292	0.3525	0.3695	0.3809
	2	0.0034	0.0125	0.0262	0.0433	0.0629	0.0840	0.1061	0.1285	0.1507
	3	0.0001	0.0006	0.0019	0.0042	0.0077	0.0125	0.0186	0.0261	0.0348
	4	0.0000	0.0000	0.0001	0.0003	0.0006	0.0012	0.0021	0.0034	0.0052
	5	0.0000	0.0000	0.0000	0.0000	0.0000	0.0001	0.0002	0.0003	0.0005
	6	0.0000	0.0000	0.0000	0.0000	0.0000	0.0000	0.0000	0.0000	0.0000
	7	0.0000	0.0000	0.0000	0.0000	0.0000	0.0000	0.0000	0.0000	0.0000
	8	0.0000	0.0000	0.0000	0.0000	0.0000	0.0000	0.0000	0.0000	0.0000
	9	0.0000	0.0000	0.0000	0.0000	0.0000	0.0000	0.0000	0.0000	0.0000
10	0	0.9044	0.8171	0.7374	0.6648	0.5987	0.5386	0.4840	0.4344	0.3894
	1	0.0914	0.1667	0.2281	0.2770	0.3151	0.3438	0.3643	0.3777	0.3851
	2	0.0042	0.0153	0.0317	0.0519	0.0746	0.0988	0.1234	0.1478	0.1714
	3	0.0001	0.0008	0.0026	0.0058	0.0105	0.0168	0.0248	0.0343	0.0452
	4	0.0000	0.0000	0.0001	0.0004	0.0010	0.0019	0.0033	0.0052	0.0078
	5	0.0000	0.0000	0.0000	0.0000	0.0001	0.0001	0.0003	0.0005	0.0009
	6	0.0000	0.0000	0.0000	0.0000	0.0000	0.0000	0.0000	0.0000	0.0001
	7	0.0000	0.0000	0.0000	0.0000	0.0000	0.0000	0.0000	0.0000	0.0000
	8	0.0000	0.0000	0.0000	0.0000	0.0000	0.0000	0.0000	0.0000	0.0000
	9	0.0000	0.0000	0.0000	0.0000	0.0000	0.0000	0.0000	0.0000	0.0000
	10	0.0000	0.0000	0.0000	0.0000	0.0000	0.0000	0.0000	0.0000	0.0000
11	0	0.8953	0.8007	0.7153	0.6382	0.5688	0.5063	0.4501	0.3996	0.3544
	1	0.0995	0.1798	0.2433	0.2925	0.3293	0.3555	0.3727	0.3823	0.3855
	2	0.0050	0.0183	0.0376	0.0609	0.0867	0.1135	0.1403	0.1662	0.1906
	3	0.0002	0.0011	0.0035	0.0076	0.0137	0.0217	0.0317	0.0434	0.0566
	4	0.0000	0.0000	0.0002	0.0006	0.0014	0.0028	0.0048	0.0075	0.0112
	5	0.0000	0.0000	0.0000	0.0000	0.0001	0.0002	0.0005	0.0009	0.0015
	6	0.0000	0.0000	0.0000	0.0000	0.0000	0.0000	0.0000	0.0001	0.0002
	7	0.0000	0.0000	0.0000	0.0000	0.0000	0.0000	0.0000	0.0000	0.0000
	8	0.0000	0.0000	0.0000	0.0000	0.0000	0.0000	0.0000	0.0000	0.0000
	9	0.0000	0.0000	0.0000	0.0000	0.0000	0.0000	0.0000	0.0000	0.0000
	10	0.0000	0.0000	0.0000	0.0000	0.0000	0.0000	0.0000	0.0000	0.0000
	11	0.0000	0.0000	0.0000	0.0000	0.0000	0.0000	0.0000	0.0000	0.0000
12	0	0.8864	0.7847	0.6938	0.6127	0.5404	0.4759	0.4186	0.3677	0.3225
	1	0.1074	0.1922	0.2575	0.3064	0.3413	0.3645	0.3781	0.3837	0.3827
	2	0.0060	0.0216	0.0438	0.0702	0.0988	0.1280	0.1565	0.1835	0.2082
	3	0.0002	0.0015	0.0045	0.0098	0.0173	0.0272	0.0393	0.0532	0.0686
	4	0.0000	0.0001	0.0003	0.0009	0.0021	0.0039	0.0067	0.0104	0.0153
	5	0.0000	0.0000	0.0000	0.0001	0.0002	0.0004	0.0008	0.0014	0.0024
	6	0.0000	0.0000	0.0000	0.0000	0.0000	0.0000	0.0001	0.0001	0.0003
	7	0.0000	0.0000	0.0000	0.0000	0.0000	0.0000	0.0000	0.0000	0.0000
	8	0.0000	0.0000	0.0000	0.0000	0.0000	0.0000	0.0000	0.0000	0.0000
	9	0.0000	0.0000	0.0000	0.0000	0.0000	0.0000	0.0000	0.0000	0.0000
	10	0.0000	0.0000	0.0000	0.0000	0.0000	0.0000	0.0000	0.0000	0.0000
	11	0.0000	0.0000	0.0000	0.0000	0.0000	0.0000	0.0000	0.0000	0.0000
	12	0.0000	0.0000	0.0000	0.0000	0.0000	0.0000	0.0000	0.0000	0.0000

Appendix 3
Table of Binomial Probabilities (Continued)

n	# of successes	p = 0.01	0.02	0.03	0.04	0.05	0.06	0.07	0.08	0.09
15	0	0.8601	0.7386	0.6333	0.5421	0.4633	0.3953	0.3367	0.2863	0.2430
	1	0.1303	0.2261	0.2938	0.3388	0.3658	0.3785	0.3801	0.3734	0.3605
	2	0.0092	0.0323	0.0636	0.0988	0.1348	0.1691	0.2003	0.2273	0.2496
	3	0.0004	0.0029	0.0085	0.0178	0.0307	0.0468	0.0653	0.0857	0.1070
	4	0.0000	0.0002	0.0008	0.0022	0.0049	0.0090	0.0148	0.0223	0.0317
	5	0.0000	0.0000	0.0001	0.0002	0.0006	0.0013	0.0024	0.0043	0.0069
	6	0.0000	0.0000	0.0000	0.0000	0.0000	0.0001	0.0003	0.0006	0.0011
	7	0.0000	0.0000	0.0000	0.0000	0.0000	0.0000	0.0000	0.0001	0.0001
	8	0.0000	0.0000	0.0000	0.0000	0.0000	0.0000	0.0000	0.0000	0.0000
	9	0.0000	0.0000	0.0000	0.0000	0.0000	0.0000	0.0000	0.0000	0.0000
	10	0.0000	0.0000	0.0000	0.0000	0.0000	0.0000	0.0000	0.0000	0.0000
	11	0.0000	0.0000	0.0000	0.0000	0.0000	0.0000	0.0000	0.0000	0.0000
	12	0.0000	0.0000	0.0000	0.0000	0.0000	0.0000	0.0000	0.0000	0.0000
	13	0.0000	0.0000	0.0000	0.0000	0.0000	0.0000	0.0000	0.0000	0.0000
	14	0.0000	0.0000	0.0000	0.0000	0.0000	0.0000	0.0000	0.0000	0.0000
	15	0.0000	0.0000	0.0000	0.0000	0.0000	0.0000	0.0000	0.0000	0.0000
20	0	0.8179	0.6676	0.5438	0.4420	0.3585	0.2901	0.2342	0.1887	0.1516
	1	0.1652	0.2725	0.3364	0.3683	0.3774	0.3703	0.3526	0.3282	0.3000
	2	0.0159	0.0528	0.0988	0.1458	0.1887	0.2246	0.2521	0.2711	0.2818
	3	0.0010	0.0065	0.0183	0.0364	0.0596	0.0860	0.1139	0.1414	0.1672
	4	0.0000	0.0006	0.0024	0.0065	0.0133	0.0233	0.0364	0.0523	0.0703
	5	0.0000	0.0000	0.0002	0.0009	0.0022	0.0048	0.0088	0.0145	0.0222
	6	0.0000	0.0000	0.0000	0.0001	0.0003	0.0008	0.0017	0.0032	0.0055
	7	0.0000	0.0000	0.0000	0.0000	0.0000	0.0001	0.0002	0.0005	0.0011
	8	0.0000	0.0000	0.0000	0.0000	0.0000	0.0000	0.0000	0.0001	0.0002
	9	0.0000	0.0000	0.0000	0.0000	0.0000	0.0000	0.0000	0.0000	0.0000
	10	0.0000	0.0000	0.0000	0.0000	0.0000	0.0000	0.0000	0.0000	0.0000
	11	0.0000	0.0000	0.0000	0.0000	0.0000	0.0000	0.0000	0.0000	0.0000
	12	0.0000	0.0000	0.0000	0.0000	0.0000	0.0000	0.0000	0.0000	0.0000
	13	0.0000	0.0000	0.0000	0.0000	0.0000	0.0000	0.0000	0.0000	0.0000
	14	0.0000	0.0000	0.0000	0.0000	0.0000	0.0000	0.0000	0.0000	0.0000
	15	0.0000	0.0000	0.0000	0.0000	0.0000	0.0000	0.0000	0.0000	0.0000
	16	0.0000	0.0000	0.0000	0.0000	0.0000	0.0000	0.0000	0.0000	0.0000
	17	0.0000	0.0000	0.0000	0.0000	0.0000	0.0000	0.0000	0.0000	0.0000
	18	0.0000	0.0000	0.0000	0.0000	0.0000	0.0000	0.0000	0.0000	0.0000
	19	0.0000	0.0000	0.0000	0.0000	0.0000	0.0000	0.0000	0.0000	0.0000
	20	0.0000	0.0000	0.0000	0.0000	0.0000	0.0000	0.0000	0.0000	0.0000

Appendix 3
Table of Binomial Probabilities (Continued)

		p								
n	# of successes	0.10	0.15	0.20	0.25	0.30	0.35	0.40	0.45	0.50
2	0	0.8100	0.7225	0.6400	0.5625	0.4900	0.4225	0.3600	0.3025	0.2500
	1	0.1800	0.2550	0.3200	0.3750	0.4200	0.4550	0.4800	0.4950	0.5000
	2	0.0100	0.0225	0.0400	0.0625	0.0900	0.1225	0.1600	0.2025	0.2500
3	0	0.7290	0.6141	0.5120	0.4219	0.3430	0.2746	0.2160	0.1664	0.1250
	1	0.2430	0.3251	0.3840	0.4219	0.4410	0.4436	0.4320	0.4084	0.3750
	2	0.0270	0.0574	0.0960	0.1406	0.1890	0.2389	0.2880	0.3341	0.3750
	3	0.0010	0.0034	0.0080	0.0156	0.0270	0.0429	0.0640	0.0911	0.1250
4	0	0.6561	0.5220	0.4096	0.3164	0.2401	0.1785	0.1296	0.0915	0.0625
	1	0.2916	0.3685	0.4096	0.4219	0.4116	0.3845	0.3456	0.2995	0.2500
	2	0.0486	0.0975	0.1536	0.2109	0.2646	0.3105	0.3456	0.3675	0.3750
	3	0.0036	0.0115	0.0256	0.0469	0.0756	0.1115	0.1536	0.2005	0.2500
	4	0.0001	0.0005	0.0016	0.0039	0.0081	0.0150	0.0256	0.0410	0.0625
5	0	0.5905	0.4437	0.3277	0.2373	0.1681	0.1160	0.0778	0.0503	0.0313
	1	0.3281	0.3915	0.4096	0.3955	0.3602	0.3124	0.2592	0.2059	0.1563
	2	0.0729	0.1382	0.2048	0.2637	0.3087	0.3364	0.3456	0.3369	0.3125
	3	0.0081	0.0244	0.0512	0.0879	0.1323	0.1811	0.2304	0.2757	0.3125
	4	0.0005	0.0022	0.0064	0.0146	0.0284	0.0488	0.0768	0.1128	0.1563
	5	0.0000	0.0001	0.0003	0.0010	0.0024	0.0053	0.0102	0.0185	0.0313
6	0	0.5314	0.3771	0.2621	0.1780	0.1176	0.0754	0.0467	0.0277	0.0156
	1	0.3543	0.3993	0.3932	0.3560	0.3025	0.2437	0.1866	0.1359	0.0938
	2	0.0984	0.1762	0.2458	0.2966	0.3241	0.3280	0.3110	0.2780	0.2344
	3	0.0146	0.0415	0.0819	0.1318	0.1852	0.2355	0.2765	0.3032	0.3125
	4	0.0012	0.0055	0.0154	0.0330	0.0595	0.0951	0.1382	0.1861	0.2344
	5	0.0001	0.0004	0.0015	0.0044	0.0102	0.0205	0.0369	0.0609	0.0938
	6	0.0000	0.0000	0.0001	0.0002	0.0007	0.0018	0.0041	0.0083	0.0156
7	0	0.4783	0.3206	0.2097	0.1335	0.0824	0.0490	0.0280	0.0152	0.0078
	1	0.3720	0.3960	0.3670	0.3115	0.2471	0.1848	0.1306	0.0872	0.0547
	2	0.1240	0.2097	0.2753	0.3115	0.3177	0.2985	0.2613	0.2140	0.1641
	3	0.0230	0.0617	0.1147	0.1730	0.2269	0.2679	0.2903	0.2918	0.2734
	4	0.0026	0.0109	0.0287	0.0577	0.0972	0.1442	0.1935	0.2388	0.2734
	5	0.0002	0.0012	0.0043	0.0115	0.0250	0.0466	0.0774	0.1172	0.1641
	6	0.0000	0.0001	0.0004	0.0013	0.0036	0.0084	0.0172	0.0320	0.0547
	7	0.0000	0.0000	0.0000	0.0001	0.0002	0.0006	0.0016	0.0037	0.0078
8	0	0.4305	0.2725	0.1678	0.1001	0.0576	0.0319	0.0168	0.0084	0.0039
	1	0.3826	0.3847	0.3355	0.2670	0.1977	0.1373	0.0896	0.0548	0.0313
	2	0.1488	0.2376	0.2936	0.3115	0.2965	0.2587	0.2090	0.1569	0.1094
	3	0.0331	0.0839	0.1468	0.2076	0.2541	0.2786	0.2787	0.2568	0.2188
	4	0.0046	0.0185	0.0459	0.0865	0.1361	0.1875	0.2322	0.2627	0.2734
	5	0.0004	0.0026	0.0092	0.0231	0.0467	0.0808	0.1239	0.1719	0.2188
	6	0.0000	0.0002	0.0011	0.0038	0.0100	0.0217	0.0413	0.0703	0.1094
	7	0.0000	0.0000	0.0001	0.0004	0.0012	0.0033	0.0079	0.0164	0.0313
	8	0.0000	0.0000	0.0000	0.0000	0.0001	0.0002	0.0007	0.0017	0.0039

Appendix 3
Table of Binomial Probabilities (Continued)

		p								
n	# of successes	0.10	0.15	0.20	0.25	0.30	0.35	0.40	0.45	0.50
9	0	0.3874	0.2316	0.1342	0.0751	0.0404	0.0207	0.0101	0.0046	0.0020
	1	0.3874	0.3679	0.3020	0.2253	0.1556	0.1004	0.0605	0.0339	0.0176
	2	0.1722	0.2597	0.3020	0.3003	0.2668	0.2162	0.1612	0.1110	0.0703
	3	0.0446	0.1069	0.1762	0.2336	0.2668	0.2716	0.2508	0.2119	0.1641
	4	0.0074	0.0283	0.0661	0.1168	0.1715	0.2194	0.2508	0.2600	0.2461
	5	0.0008	0.0050	0.0165	0.0389	0.0735	0.1181	0.1672	0.2128	0.2461
	6	0.0001	0.0006	0.0028	0.0087	0.0210	0.0424	0.0743	0.1160	0.1641
	7	0.0000	0.0000	0.0003	0.0012	0.0039	0.0098	0.0212	0.0407	0.0703
	8	0.0000	0.0000	0.0000	0.0001	0.0004	0.0013	0.0035	0.0083	0.0176
	9	0.0000	0.0000	0.0000	0.0000	0.0000	0.0001	0.0003	0.0008	0.0020
10	0	0.3487	0.1969	0.1074	0.0563	0.0282	0.0135	0.0060	0.0025	0.0010
	1	0.3874	0.3474	0.2684	0.1877	0.1211	0.0725	0.0403	0.0207	0.0098
	2	0.1937	0.2759	0.3020	0.2816	0.2335	0.1757	0.1209	0.0763	0.0439
	3	0.0574	0.1298	0.2013	0.2503	0.2668	0.2522	0.2150	0.1665	0.1172
	4	0.0112	0.0401	0.0881	0.1460	0.2001	0.2377	0.2508	0.2384	0.2051
	5	0.0015	0.0085	0.0264	0.0584	0.1029	0.1536	0.2007	0.2340	0.2461
	6	0.0001	0.0012	0.0055	0.0162	0.0368	0.0689	0.1115	0.1596	0.2051
	7	0.0000	0.0001	0.0008	0.0031	0.0090	0.0212	0.0425	0.0746	0.1172
	8	0.0000	0.0000	0.0001	0.0004	0.0014	0.0043	0.0106	0.0229	0.0439
	9	0.0000	0.0000	0.0000	0.0000	0.0001	0.0005	0.0016	0.0042	0.0098
	10	0.0000	0.0000	0.0000	0.0000	0.0000	0.0000	0.0001	0.0003	0.0010
11	0	0.3138	0.1673	0.0859	0.0422	0.0198	0.0088	0.0036	0.0014	0.0005
	1	0.3835	0.3248	0.2362	0.1549	0.0932	0.0518	0.0266	0.0125	0.0054
	2	0.2131	0.2866	0.2953	0.2581	0.1998	0.1395	0.0887	0.0513	0.0269
	3	0.0710	0.1517	0.2215	0.2581	0.2568	0.2254	0.1774	0.1259	0.0806
	4	0.0158	0.0536	0.1107	0.1721	0.2201	0.2428	0.2365	0.2060	0.1611
	5	0.0025	0.0132	0.0388	0.0803	0.1321	0.1830	0.2207	0.2360	0.2256
	6	0.0003	0.0023	0.0097	0.0268	0.0566	0.0985	0.1471	0.1931	0.2256
	7	0.0000	0.0003	0.0017	0.0064	0.0173	0.0379	0.0701	0.1128	0.1611
	8	0.0000	0.0000	0.0002	0.0011	0.0037	0.0102	0.0234	0.0462	0.0806
	9	0.0000	0.0000	0.0000	0.0001	0.0005	0.0018	0.0052	0.0126	0.0269
	10	0.0000	0.0000	0.0000	0.0000	0.0000	0.0002	0.0007	0.0021	0.0054
	11	0.0000	0.0000	0.0000	0.0000	0.0000	0.0000	0.0000	0.0002	0.0005
12	0	0.2824	0.1422	0.0687	0.0317	0.0138	0.0057	0.0022	0.0008	0.0002
	1	0.3766	0.3012	0.2062	0.1267	0.0712	0.0368	0.0174	0.0075	0.0029
	2	0.2301	0.2924	0.2835	0.2323	0.1678	0.1088	0.0639	0.0339	0.0161
	3	0.0852	0.1720	0.2362	0.2581	0.2397	0.1954	0.1419	0.0923	0.0537
	4	0.0213	0.0683	0.1329	0.1936	0.2311	0.2367	0.2128	0.1700	0.1208
	5	0.0038	0.0193	0.0532	0.1032	0.1585	0.2039	0.2270	0.2225	0.1934
	6	0.0005	0.0040	0.0155	0.0401	0.0792	0.1281	0.1766	0.2124	0.2256
	7	0.0000	0.0006	0.0033	0.0115	0.0291	0.0591	0.1009	0.1489	0.1934
	8	0.0000	0.0001	0.0005	0.0024	0.0078	0.0199	0.0420	0.0762	0.1208
	9	0.0000	0.0000	0.0001	0.0004	0.0015	0.0048	0.0125	0.0277	0.0537
	10	0.0000	0.0000	0.0000	0.0000	0.0002	0.0008	0.0025	0.0068	0.0161
	11	0.0000	0.0000	0.0000	0.0000	0.0000	0.0001	0.0003	0.0010	0.0029
	12	0.0000	0.0000	0.0000	0.0000	0.0000	0.0000	0.0000	0.0001	0.0002

Appendix 3
Table of Binomial Probabilities (Continued)

		p								
n	# of successes	0.10	0.15	0.20	0.25	0.30	0.35	0.40	0.45	0.50
15	0	0.2059	0.0874	0.0352	0.0134	0.0047	0.0016	0.0005	0.0001	0.0000
	1	0.3432	0.2312	0.1319	0.0668	0.0305	0.0126	0.0047	0.0016	0.0005
	2	0.2669	0.2856	0.2309	0.1559	0.0916	0.0476	0.0219	0.0090	0.0032
	3	0.1285	0.2184	0.2501	0.2252	0.1700	0.1110	0.0634	0.0318	0.0139
	4	0.0428	0.1156	0.1876	0.2252	0.2186	0.1792	0.1268	0.0780	0.0417
	5	0.0105	0.0449	0.1032	0.1651	0.2061	0.2123	0.1859	0.1404	0.0916
	6	0.0019	0.0132	0.0430	0.0917	0.1472	0.1906	0.2066	0.1914	0.1527
	7	0.0003	0.0030	0.0138	0.0393	0.0811	0.1319	0.1771	0.2013	0.1964
	8	0.0000	0.0005	0.0035	0.0131	0.0348	0.0710	0.1181	0.1647	0.1964
	9	0.0000	0.0001	0.0007	0.0034	0.0116	0.0298	0.0612	0.1048	0.1527
	10	0.0000	0.0000	0.0001	0.0007	0.0030	0.0096	0.0245	0.0515	0.0916
	11	0.0000	0.0000	0.0000	0.0001	0.0006	0.0024	0.0074	0.0191	0.0417
	12	0.0000	0.0000	0.0000	0.0000	0.0001	0.0004	0.0016	0.0052	0.0139
	13	0.0000	0.0000	0.0000	0.0000	0.0000	0.0001	0.0003	0.0010	0.0032
	14	0.0000	0.0000	0.0000	0.0000	0.0000	0.0000	0.0000	0.0001	0.0005
	15	0.0000	0.0000	0.0000	0.0000	0.0000	0.0000	0.0000	0.0000	0.0000
20	0	0.1216	0.0388	0.0115	0.0032	0.0008	0.0002	0.0000	0.0000	0.0000
	1	0.2702	0.1368	0.0576	0.0211	0.0068	0.0020	0.0005	0.0001	0.0000
	2	0.2852	0.2293	0.1369	0.0669	0.0278	0.0100	0.0031	0.0008	0.0002
	3	0.1901	0.2428	0.2054	0.1339	0.0716	0.0323	0.0123	0.0040	0.0011
	4	0.0898	0.1821	0.2182	0.1897	0.1304	0.0738	0.0350	0.0139	0.0046
	5	0.0319	0.1028	0.1746	0.2023	0.1789	0.1272	0.0746	0.0365	0.0148
	6	0.0089	0.0454	0.1091	0.1686	0.1916	0.1712	0.1244	0.0746	0.0370
	7	0.0020	0.0160	0.0545	0.1124	0.1643	0.1844	0.1659	0.1221	0.0739
	8	0.0004	0.0046	0.0222	0.0609	0.1144	0.1614	0.1797	0.1623	0.1201
	9	0.0001	0.0011	0.0074	0.0271	0.0654	0.1158	0.1597	0.1771	0.1602
	10	0.0000	0.0002	0.0020	0.0099	0.0308	0.0686	0.1171	0.1593	0.1762
	11	0.0000	0.0000	0.0005	0.0030	0.0120	0.0336	0.0710	0.1185	0.1602
	12	0.0000	0.0000	0.0001	0.0008	0.0039	0.0136	0.0355	0.0727	0.1201
	13	0.0000	0.0000	0.0000	0.0002	0.0010	0.0045	0.0146	0.0366	0.0739
	14	0.0000	0.0000	0.0000	0.0000	0.0002	0.0012	0.0049	0.0150	0.0370
	15	0.0000	0.0000	0.0000	0.0000	0.0000	0.0003	0.0013	0.0049	0.0148
	16	0.0000	0.0000	0.0000	0.0000	0.0000	0.0000	0.0003	0.0013	0.0046
	17	0.0000	0.0000	0.0000	0.0000	0.0000	0.0000	0.0000	0.0002	0.0011
	18	0.0000	0.0000	0.0000	0.0000	0.0000	0.0000	0.0000	0.0000	0.0002
	19	0.0000	0.0000	0.0000	0.0000	0.0000	0.0000	0.0000	0.0000	0.0000
	20	0.0000	0.0000	0.0000	0.0000	0.0000	0.0000	0.0000	0.0000	0.0000

Appendix 4
Critical Values of the Chi Square Distribution
(Chi Square is inherently a two-sided test)

Degrees of Freedom	α = Level of Significance					
ν	0.200	0.100	0.050	0.025	0.010	0.005
1	1.6424	2.706	3.841	5.024	6.635	7.879
2	3.2189	4.605	5.991	7.378	9.210	10.60
3	4.6416	6.251	7.815	9.348	11.34	12.84
4	5.989	7.779	9.488	11.14	13.28	14.86
5	7.289	9.236	11.07	12.83	15.09	16.75
6	8.558	10.64	12.59	14.45	16.81	18.55
7	9.803	12.02	14.07	16.01	18.48	20.28
8	11.030	13.36	15.51	17.53	20.09	21.95
9	12.242	14.68	16.92	19.02	21.67	23.59
10	13.442	15.99	18.31	20.48	23.21	25.19
11	14.631	17.28	19.68	21.92	24.73	26.76
12	15.812	18.55	21.03	23.34	26.22	28.30
13	16.985	19.81	22.36	24.74	27.69	29.82
14	18.151	21.06	23.68	26.12	29.14	31.32
15	19.311	22.31	25.00	27.49	30.58	32.80
16	20.465	23.54	26.30	28.85	32.00	34.27
17	21.61	24.77	27.59	30.19	33.41	35.72
18	22.76	25.99	28.87	31.53	34.81	37.16
19	23.90	27.20	30.14	32.85	36.19	38.58
20	25.04	28.41	31.41	34.17	37.57	40.00
21	26.17	29.62	32.67	35.48	38.93	41.40
22	27.30	30.81	33.92	36.78	40.29	42.80
23	28.43	32.01	35.17	38.08	41.64	44.18
24	29.55	33.20	36.42	39.36	42.98	45.56
25	30.68	34.38	37.65	40.65	44.31	46.93
26	31.79	35.56	38.89	41.92	45.64	48.29
27	32.91	36.74	40.11	43.19	46.96	49.65
28	34.03	37.92	41.34	44.46	48.28	50.99
29	35.14	39.09	42.56	45.72	49.59	52.34
30	36.25	40.26	43.77	46.98	50.89	53.67
35	41.78	46.06	49.80	53.20	57.34	60.27
40	47.27	51.81	55.76	59.34	63.69	66.77
45	52.73	57.51	61.66	65.41	69.96	73.17
50	58.16	63.17	67.50	71.42	76.15	79.49
55	63.58	68.80	73.31	77.38	82.29	85.75
60	68.97	74.40	79.08	83.30	88.38	91.95
65	74.35	79.97	84.82	89.18	94.42	98.10
70	79.71	85.53	90.53	95.02	100.4	104.2
75	85.07	91.06	96.22	100.8	106.4	110.3
80	90.41	96.58	101.9	106.6	112.3	116.3
85	95.73	102.1	107.5	112.4	118.2	122.3
90	101.05	107.6	113.1	118.1	124.1	128.3
95	106.36	113.0	118.8	123.9	130.0	134.2
100	111.67	118.5	124.3	129.6	135.8	140.2

Appendix 5
Binomial Probabilities for Performing the Sign Test
(The Sign Test is inherently one-sided, for a two-sided test double the corresponding table values)

n \ x	0	1	2	3	4	5	6	7	8	9	10
1	0.5000	*									
2	0.2500	0.5000	*								
3	0.1250	0.5000	0.8750	*							
4	0.0625	0.3125	0.6875	0.9375	*						
5	0.0313	0.1875	0.5000	0.8125	0.9688	*					
6	0.0156	0.1094	0.3438	0.6563	0.8906	0.9844	*				
7	0.0078	0.0625	0.2266	0.5000	0.7734	0.9375	0.9922	*			
8	0.0039	0.0352	0.1445	0.3633	0.6367	0.8555	0.9648	0.9961	*		
9	0.0020	0.0195	0.0898	0.2539	0.5000	0.7461	0.9102	0.9805	0.9980	*	
10	0.0010	0.0107	0.0547	0.1719	0.3770	0.6230	0.8281	0.9453	0.9893	0.9990	*
11	0.0005	0.0059	0.0327	0.1133	0.2744	0.5000	0.7256	0.8867	0.9673	0.9941	0.9995
12	0.0002	0.0032	0.0193	0.0730	0.1938	0.3872	0.6128	0.8062	0.9270	0.9807	0.9968
13	0.0001	0.0017	0.0112	0.0461	0.1334	0.2905	0.5000	0.7095	0.8666	0.9539	0.9888
14	0.0001	0.0009	0.0065	0.0287	0.0898	0.2120	0.3953	0.6047	0.7880	0.9102	0.9713
15	0.0000	0.0005	0.0037	0.0176	0.0592	0.1509	0.3036	0.5000	0.6964	0.8491	0.9408
16	0.0000	0.0003	0.0021	0.0106	0.0384	0.1051	0.2272	0.4018	0.5982	0.7728	0.8949
17	0.0000	0.0001	0.0012	0.0064	0.0245	0.0717	0.1662	0.3145	0.5000	0.6855	0.8338
18	0.0000	0.0001	0.0007	0.0038	0.0154	0.0481	0.1189	0.2403	0.4073	0.5927	0.7597
19	0.0000	0.0000	0.0004	0.0022	0.0096	0.0318	0.0835	0.1796	0.3238	0.5000	0.6762
20	0.0000	0.0000	0.0002	0.0013	0.0059	0.0207	0.0577	0.1316	0.2517	0.4119	0.5881

One-sided probabilities for values less than or equal to "x" from "n" trials

* Approximately 1.0

Appendix 6
Critical Values for the Wilcoxon Signed-Rank Test

	Level of Significance, α, for a one-tail test			
	0.05	0.025	0.01	0.005
	Level of Significance, α, for a two-tail test			
n	0.1	0.05	0.02	0.01
6	2	1	-	-
7	4	2	0	-
8	6	4	2	0
9	8	6	3	2
10	11	8	5	3
11	14	11	7	5
12	17	14	10	7
13	21	17	13	10
14	26	21	16	13
15	30	25	20	16
16	36	30	24	19
17	41	35	28	23
18	47	40	33	28
19	54	46	38	32
20	60	52	43	37
21	68	59	49	43
22	75	66	56	49
23	83	73	62	55
24	92	81	69	61
25	101	90	77	68

Appendix 7
Student's t Distribution

degrees of freedom	.10	.05	.025	.0125	.010	.005	one-tailed probabilities
	.20	.10	.050	.025	.020	.010	two-tailed probabilities
1	3.078	6.314	12.706	25.452	31.821	63.656	
2	1.886	2.920	4.303	6.205	6.965	9.925	
3	1.638	2.353	3.182	4.177	4.541	5.841	
4	1.533	2.132	2.776	3.495	3.747	4.604	
5	1.476	2.015	2.571	3.163	3.365	4.032	
6	1.440	1.943	2.447	2.969	3.143	3.707	
7	1.415	1.895	2.365	2.841	2.998	3.499	
8	1.397	1.860	2.306	2.752	2.896	3.355	
9	1.383	1.833	2.262	2.685	2.821	3.250	
10	1.372	1.812	2.228	2.634	2.764	3.169	
11	1.363	1.796	2.201	2.593	2.718	3.106	
12	1.356	1.782	2.179	2.560	2.681	3.055	
13	1.350	1.771	2.160	2.533	2.650	3.012	
14	1.345	1.761	2.145	2.510	2.624	2.977	
15	1.341	1.753	2.131	2.490	2.602	2.947	
16	1.337	1.746	2.120	2.473	2.583	2.921	
17	1.333	1.740	2.110	2.458	2.567	2.898	
18	1.330	1.734	2.101	2.445	2.552	2.878	
19	1.328	1.729	2.093	2.433	2.539	2.861	
20	1.325	1.725	2.086	2.423	2.528	2.845	
21	1.323	1.721	2.080	2.414	2.518	2.831	
22	1.321	1.717	2.074	2.405	2.508	2.819	
23	1.319	1.714	2.069	2.398	2.500	2.807	
24	1.318	1.711	2.064	2.391	2.492	2.797	
25	1.316	1.708	2.060	2.385	2.485	2.787	
26	1.315	1.706	2.056	2.379	2.479	2.779	
27	1.314	1.703	2.052	2.373	2.473	2.771	
28	1.313	1.701	2.048	2.368	2.467	2.763	
29	1.311	1.699	2.045	2.364	2.462	2.756	
30	1.310	1.697	2.042	2.360	2.457	2.750	
40	1.303	1.684	2.021	2.329	2.423	2.704	
70	1.294	1.667	1.994	2.291	2.381	2.648	
100	1.290	1.660	1.984	2.276	2.364	2.626	

Appendix 8
F Ratio Critical Values

$\alpha = 0.10$

d.f. denominator	d.f. numerator 1	2	3	4	5	6	7	8	9	10	11	12	13	14	15	16
1	39.86	49.50	53.59	55.83	57.24	58.20	58.91	59.44	59.86	60.19	60.47	60.71	60.90	61.07	61.22	61.35
2	8.53	9.00	9.16	9.24	9.29	9.33	9.35	9.37	9.38	9.39	9.40	9.41	9.41	9.42	9.42	9.43
3	5.54	5.46	5.39	5.34	5.31	5.28	5.27	5.25	5.24	5.23	5.22	5.22	5.21	5.20	5.20	5.20
4	4.54	4.32	4.19	4.11	4.05	4.01	3.98	3.95	3.94	3.92	3.91	3.90	3.89	3.88	3.87	3.86
5	4.06	3.78	3.62	3.52	3.45	3.40	3.37	3.34	3.32	3.30	3.28	3.27	3.26	3.25	3.24	3.23
6	3.78	3.46	3.29	3.18	3.11	3.05	3.01	2.98	2.96	2.94	2.92	2.90	2.89	2.88	2.87	2.86
7	3.59	3.26	3.07	2.96	2.88	2.83	2.78	2.75	2.72	2.70	2.68	2.67	2.65	2.64	2.63	2.62
8	3.46	3.11	2.92	2.81	2.73	2.67	2.62	2.59	2.56	2.54	2.52	2.50	2.49	2.48	2.46	2.45
9	3.36	3.01	2.81	2.69	2.61	2.55	2.51	2.47	2.44	2.42	2.40	2.38	2.36	2.35	2.34	2.33
10	3.29	2.92	2.73	2.61	2.52	2.46	2.41	2.38	2.35	2.32	2.30	2.28	2.27	2.26	2.24	2.23
11	3.23	2.86	2.66	2.54	2.45	2.39	2.34	2.30	2.27	2.25	2.23	2.21	2.19	2.18	2.17	2.16
12	3.18	2.81	2.61	2.48	2.39	2.33	2.28	2.24	2.21	2.19	2.17	2.15	2.13	2.12	2.10	2.09
13	3.14	2.76	2.56	2.43	2.35	2.28	2.23	2.20	2.16	2.14	2.12	2.10	2.08	2.07	2.05	2.04
14	3.10	2.73	2.52	2.39	2.31	2.24	2.19	2.15	2.12	2.10	2.07	2.05	2.04	2.02	2.01	2.00
15	3.07	2.70	2.49	2.36	2.27	2.21	2.16	2.12	2.09	2.06	2.04	2.02	2.00	1.99	1.97	1.96
16	3.05	2.67	2.46	2.33	2.24	2.18	2.13	2.09	2.06	2.03	2.01	1.99	1.97	1.95	1.94	1.93
17	3.03	2.64	2.44	2.31	2.22	2.15	2.10	2.06	2.03	2.00	1.98	1.96	1.94	1.93	1.91	1.90
18	3.01	2.62	2.42	2.29	2.20	2.13	2.08	2.04	2.00	1.98	1.95	1.93	1.92	1.90	1.89	1.87
19	2.99	2.61	2.40	2.27	2.18	2.11	2.06	2.02	1.98	1.96	1.93	1.91	1.89	1.88	1.86	1.85
20	2.97	2.59	2.38	2.25	2.16	2.09	2.04	2.00	1.96	1.94	1.91	1.89	1.87	1.86	1.84	1.83
22	2.95	2.56	2.35	2.22	2.13	2.06	2.01	1.97	1.93	1.90	1.88	1.86	1.84	1.83	1.81	1.80
23	2.94	2.55	2.34	2.21	2.11	2.05	1.99	1.95	1.92	1.89	1.87	1.84	1.83	1.81	1.80	1.78
24	2.93	2.54	2.33	2.19	2.10	2.04	1.98	1.94	1.91	1.88	1.85	1.83	1.81	1.80	1.78	1.77
25	2.92	2.53	2.32	2.18	2.09	2.02	1.97	1.93	1.89	1.87	1.84	1.82	1.80	1.79	1.77	1.76
30	2.88	2.49	2.28	2.14	2.05	1.98	1.93	1.88	1.85	1.82	1.79	1.77	1.75	1.74	1.72	1.71
40	2.84	2.44	2.23	2.09	2.00	1.93	1.87	1.83	1.79	1.76	1.74	1.71	1.70	1.68	1.66	1.65
50	2.81	2.41	2.20	2.06	1.97	1.90	1.84	1.80	1.76	1.73	1.70	1.68	1.66	1.64	1.63	1.61
60	2.79	2.39	2.18	2.04	1.95	1.87	1.82	1.77	1.74	1.71	1.68	1.66	1.64	1.62	1.60	1.59
75	2.77	2.37	2.16	2.02	1.93	1.85	1.80	1.75	1.72	1.69	1.66	1.63	1.61	1.60	1.58	1.57
100	2.76	2.36	2.14	2.00	1.91	1.83	1.78	1.73	1.69	1.66	1.64	1.61	1.59	1.57	1.56	1.54
200	2.73	2.33	2.11	1.97	1.88	1.80	1.75	1.70	1.66	1.63	1.60	1.58	1.56	1.54	1.52	1.51
1000	2.71	2.31	2.09	1.95	1.85	1.78	1.72	1.68	1.64	1.61	1.58	1.55	1.53	1.51	1.49	1.48

Appendix 8
F Ratio Critical Values

$\alpha = 0.10$

d.f. denominator	d.f. numerator 17	18	19	20	21	22	23	24	25	30	40	50	60	75	100	200	1000
1	61.46	61.57	61.66	61.74	61.81	61.88	61.94	62.00	62.05	62.26	62.53	62.69	62.79	62.90	63.01	63.17	63.30
2	9.43	9.44	9.44	9.44	9.44	9.45	9.45	9.45	9.45	9.46	9.47	9.47	9.47	9.48	9.48	9.49	9.49
3	5.19	5.19	5.19	5.18	5.18	5.18	5.18	5.18	5.17	5.17	5.16	5.15	5.15	5.15	5.14	5.14	5.13
4	3.86	3.85	3.85	3.84	3.84	3.84	3.83	3.83	3.83	3.82	3.80	3.80	3.79	3.78	3.78	3.77	3.76
5	3.22	3.22	3.21	3.21	3.20	3.20	3.19	3.19	3.19	3.17	3.16	3.15	3.14	3.13	3.13	3.12	3.11
6	2.85	2.85	2.84	2.84	2.83	2.83	2.82	2.82	2.81	2.80	2.78	2.77	2.76	2.75	2.75	2.73	2.72
7	2.61	2.61	2.60	2.59	2.59	2.58	2.58	2.58	2.57	2.56	2.54	2.52	2.51	2.51	2.50	2.48	2.47
8	2.45	2.44	2.43	2.42	2.42	2.41	2.41	2.40	2.40	2.38	2.36	2.35	2.34	2.33	2.32	2.31	2.30
9	2.32	2.31	2.30	2.30	2.29	2.29	2.28	2.28	2.27	2.25	2.23	2.22	2.21	2.20	2.19	2.17	2.16
10	2.22	2.22	2.21	2.20	2.19	2.19	2.18	2.18	2.17	2.16	2.13	2.12	2.11	2.10	2.09	2.07	2.06
11	2.15	2.14	2.13	2.12	2.12	2.11	2.11	2.10	2.10	2.08	2.05	2.04	2.03	2.02	2.01	1.99	1.98
12	2.08	2.08	2.07	2.06	2.05	2.05	2.04	2.04	2.03	2.01	1.99	1.97	1.96	1.95	1.94	1.92	1.91
13	2.03	2.02	2.01	2.01	2.00	1.99	1.99	1.98	1.98	1.96	1.93	1.92	1.90	1.89	1.88	1.86	1.85
14	1.99	1.98	1.97	1.96	1.96	1.95	1.94	1.94	1.93	1.91	1.89	1.87	1.86	1.85	1.83	1.82	1.80
15	1.95	1.94	1.93	1.92	1.92	1.91	1.90	1.90	1.89	1.87	1.85	1.83	1.82	1.80	1.79	1.77	1.76
16	1.92	1.91	1.90	1.89	1.88	1.88	1.87	1.87	1.86	1.84	1.81	1.79	1.78	1.77	1.76	1.74	1.72
17	1.89	1.88	1.87	1.86	1.86	1.85	1.84	1.84	1.83	1.81	1.78	1.76	1.75	1.74	1.73	1.71	1.69
18	1.86	1.85	1.84	1.84	1.83	1.82	1.82	1.81	1.80	1.78	1.75	1.74	1.72	1.71	1.70	1.68	1.66
19	1.84	1.83	1.82	1.81	1.81	1.80	1.79	1.79	1.78	1.76	1.73	1.71	1.70	1.69	1.67	1.65	1.64
20	1.82	1.81	1.80	1.79	1.79	1.78	1.77	1.77	1.76	1.74	1.71	1.69	1.68	1.66	1.65	1.63	1.61
22	1.79	1.78	1.77	1.76	1.75	1.74	1.74	1.73	1.73	1.70	1.67	1.65	1.64	1.63	1.61	1.59	1.57
23	1.77	1.76	1.75	1.74	1.74	1.73	1.72	1.72	1.71	1.69	1.66	1.64	1.62	1.61	1.59	1.57	1.55
24	1.76	1.75	1.74	1.73	1.72	1.71	1.71	1.70	1.70	1.67	1.64	1.62	1.61	1.59	1.58	1.56	1.54
25	1.75	1.74	1.73	1.72	1.71	1.70	1.70	1.69	1.68	1.66	1.63	1.61	1.59	1.58	1.56	1.54	1.52
30	1.70	1.69	1.68	1.67	1.66	1.65	1.64	1.64	1.63	1.61	1.57	1.55	1.54	1.52	1.51	1.48	1.46
40	1.64	1.62	1.61	1.61	1.60	1.59	1.58	1.57	1.57	1.54	1.51	1.48	1.47	1.45	1.43	1.41	1.38
50	1.60	1.59	1.58	1.57	1.56	1.55	1.54	1.54	1.53	1.50	1.46	1.44	1.42	1.41	1.39	1.36	1.33
60	1.58	1.56	1.55	1.54	1.53	1.53	1.52	1.51	1.50	1.48	1.44	1.41	1.40	1.38	1.36	1.33	1.30
75	1.55	1.54	1.53	1.52	1.51	1.50	1.49	1.49	1.48	1.45	1.41	1.38	1.37	1.35	1.33	1.29	1.26
100	1.53	1.52	1.50	1.49	1.48	1.48	1.47	1.46	1.45	1.42	1.38	1.35	1.34	1.32	1.29	1.26	1.22
200	1.49	1.48	1.47	1.46	1.45	1.44	1.43	1.42	1.41	1.38	1.34	1.31	1.29	1.27	1.24	1.20	1.16
1000	1.46	1.45	1.44	1.43	1.42	1.41	1.40	1.39	1.38	1.35	1.30	1.27	1.25	1.23	1.20	1.15	1.08

Appendix 8
F Ratio Critical Values

$\alpha = 0.05$

d.f. denominator	d.f. numerator 1	2	3	4	5	6	7	8	9	10	11	12	13	14	15	16
1	161.45	199.50	215.71	224.58	230.16	233.99	236.77	238.88	240.54	241.88	242.98	243.90	244.69	245.36	245.95	246.47
2	18.51	19.00	19.16	19.25	19.30	19.33	19.35	19.37	19.38	19.40	19.40	19.41	19.42	19.42	19.43	19.43
3	10.13	9.55	9.28	9.12	9.01	8.94	8.89	8.85	8.81	8.79	8.76	8.74	8.73	8.71	8.70	8.69
4	7.71	6.94	6.59	6.39	6.26	6.16	6.09	6.04	6.00	5.96	5.94	5.91	5.89	5.87	5.86	5.84
5	6.61	5.79	5.41	5.19	5.05	4.95	4.88	4.82	4.77	4.74	4.70	4.68	4.66	4.64	4.62	4.60
6	5.99	5.14	4.76	4.53	4.39	4.28	4.21	4.15	4.10	4.06	4.03	4.00	3.98	3.96	3.94	3.92
7	5.59	4.74	4.35	4.12	3.97	3.87	3.79	3.73	3.68	3.64	3.60	3.57	3.55	3.53	3.51	3.49
8	5.32	4.46	4.07	3.84	3.69	3.58	3.50	3.44	3.39	3.35	3.31	3.28	3.26	3.24	3.22	3.20
9	5.12	4.26	3.86	3.63	3.48	3.37	3.29	3.23	3.18	3.14	3.10	3.07	3.05	3.03	3.01	2.99
10	4.96	4.10	3.71	3.48	3.33	3.22	3.14	3.07	3.02	2.98	2.94	2.91	2.89	2.86	2.85	2.83
11	4.84	3.98	3.59	3.36	3.20	3.09	3.01	2.95	2.90	2.85	2.82	2.79	2.76	2.74	2.72	2.70
12	4.75	3.89	3.49	3.26	3.11	3.00	2.91	2.85	2.80	2.75	2.72	2.69	2.66	2.64	2.62	2.60
13	4.67	3.81	3.41	3.18	3.03	2.92	2.83	2.77	2.71	2.67	2.63	2.60	2.58	2.55	2.53	2.51
14	4.60	3.74	3.34	3.11	2.96	2.85	2.76	2.70	2.65	2.60	2.57	2.53	2.51	2.48	2.46	2.44
15	4.54	3.68	3.29	3.06	2.90	2.79	2.71	2.64	2.59	2.54	2.51	2.48	2.45	2.42	2.40	2.38
16	4.49	3.63	3.24	3.01	2.85	2.74	2.66	2.59	2.54	2.49	2.46	2.42	2.40	2.37	2.35	2.33
17	4.45	3.59	3.20	2.96	2.81	2.70	2.61	2.55	2.49	2.45	2.41	2.38	2.35	2.33	2.31	2.29
18	4.41	3.55	3.16	2.93	2.77	2.66	2.58	2.51	2.46	2.41	2.37	2.34	2.31	2.29	2.27	2.25
19	4.38	3.52	3.13	2.90	2.74	2.63	2.54	2.48	2.42	2.38	2.34	2.31	2.28	2.26	2.23	2.21
20	4.35	3.49	3.10	2.87	2.71	2.60	2.51	2.45	2.39	2.35	2.31	2.28	2.25	2.22	2.20	2.18
22	4.30	3.44	3.05	2.82	2.66	2.55	2.46	2.40	2.34	2.30	2.26	2.23	2.20	2.17	2.15	2.13
23	4.28	3.42	3.03	2.80	2.64	2.53	2.44	2.37	2.32	2.27	2.24	2.20	2.18	2.15	2.13	2.11
24	4.26	3.40	3.01	2.78	2.62	2.51	2.42	2.36	2.30	2.25	2.22	2.18	2.15	2.13	2.11	2.09
25	4.24	3.39	2.99	2.76	2.60	2.49	2.40	2.34	2.28	2.24	2.20	2.16	2.14	2.11	2.09	2.07
30	4.17	3.32	2.92	2.69	2.53	2.42	2.33	2.27	2.21	2.16	2.13	2.09	2.06	2.04	2.01	1.99
40	4.08	3.23	2.84	2.61	2.45	2.34	2.25	2.18	2.12	2.08	2.04	2.00	1.97	1.95	1.92	1.90
50	4.03	3.18	2.79	2.56	2.40	2.29	2.20	2.13	2.07	2.03	1.99	1.95	1.92	1.89	1.87	1.85
60	4.00	3.15	2.76	2.53	2.37	2.25	2.17	2.10	2.04	1.99	1.95	1.92	1.89	1.86	1.84	1.82
75	3.97	3.12	2.73	2.49	2.34	2.22	2.13	2.06	2.01	1.96	1.92	1.88	1.85	1.83	1.80	1.78
100	3.94	3.09	2.70	2.46	2.31	2.19	2.10	2.03	1.97	1.93	1.89	1.85	1.82	1.79	1.77	1.75
200	3.89	3.04	2.65	2.42	2.26	2.14	2.06	1.98	1.93	1.88	1.84	1.80	1.77	1.74	1.72	1.69
1000	3.85	3.00	2.61	2.38	2.22	2.11	2.02	1.95	1.89	1.84	1.80	1.76	1.73	1.70	1.68	1.65

Appendix 8
F Ratio Critical Values

α = 0.05

d.f. denominator	d.f. numerator 17	18	19	20	21	22	23	24	25	30	40	50	60	75	100	200	1000
1	246.92	247.32	247.69	248.02	248.31	248.58	248.82	249.05	249.26	250.10	251.14	251.77	252.20	252.62	253.04	253.68	254.19
2	19.44	19.44	19.44	19.45	19.45	19.45	19.45	19.45	19.46	19.46	19.47	19.48	19.48	19.48	19.49	19.49	19.49
3	8.68	8.67	8.67	8.66	8.65	8.65	8.64	8.64	8.63	8.62	8.59	8.58	8.57	8.56	8.55	8.54	8.53
4	5.83	5.82	5.81	5.80	5.79	5.79	5.78	5.77	5.77	5.75	5.72	5.70	5.69	5.68	5.66	5.65	5.63
5	4.59	4.58	4.57	4.56	4.55	4.54	4.53	4.53	4.52	4.50	4.46	4.44	4.43	4.42	4.41	4.39	4.37
6	3.91	3.90	3.88	3.87	3.86	3.86	3.85	3.84	3.83	3.81	3.77	3.75	3.74	3.73	3.71	3.69	3.67
7	3.48	3.47	3.46	3.44	3.43	3.43	3.42	3.41	3.40	3.38	3.34	3.32	3.30	3.29	3.27	3.25	3.23
8	3.19	3.17	3.16	3.15	3.14	3.13	3.12	3.12	3.11	3.08	3.04	3.02	3.01	2.99	2.97	2.95	2.93
9	2.97	2.96	2.95	2.94	2.93	2.92	2.91	2.90	2.89	2.86	2.83	2.80	2.79	2.77	2.76	2.73	2.71
10	2.81	2.80	2.79	2.77	2.76	2.75	2.75	2.74	2.73	2.70	2.66	2.64	2.62	2.60	2.59	2.56	2.54
11	2.69	2.67	2.66	2.65	2.64	2.63	2.62	2.61	2.60	2.57	2.53	2.51	2.49	2.47	2.46	2.43	2.41
12	2.58	2.57	2.56	2.54	2.53	2.52	2.51	2.51	2.50	2.47	2.43	2.40	2.38	2.37	2.35	2.32	2.30
13	2.50	2.48	2.47	2.46	2.45	2.44	2.43	2.42	2.41	2.38	2.34	2.31	2.30	2.28	2.26	2.23	2.21
14	2.43	2.41	2.40	2.39	2.38	2.37	2.36	2.35	2.34	2.31	2.27	2.24	2.22	2.21	2.19	2.16	2.14
15	2.37	2.35	2.34	2.33	2.32	2.31	2.30	2.29	2.28	2.25	2.20	2.18	2.16	2.14	2.12	2.10	2.07
16	2.32	2.30	2.29	2.28	2.26	2.25	2.24	2.24	2.23	2.19	2.15	2.12	2.11	2.09	2.07	2.04	2.02
17	2.27	2.26	2.24	2.23	2.22	2.21	2.20	2.19	2.18	2.15	2.10	2.08	2.06	2.04	2.02	1.99	1.97
18	2.23	2.22	2.20	2.19	2.18	2.17	2.16	2.15	2.14	2.11	2.06	2.04	2.02	2.00	1.98	1.95	1.92
19	2.20	2.18	2.17	2.16	2.14	2.13	2.12	2.11	2.11	2.07	2.03	2.00	1.98	1.96	1.94	1.91	1.88
20	2.17	2.15	2.14	2.12	2.11	2.10	2.09	2.08	2.07	2.04	1.99	1.97	1.95	1.93	1.91	1.88	1.85
22	2.11	2.10	2.08	2.07	2.06	2.05	2.04	2.03	2.02	1.98	1.94	1.91	1.89	1.87	1.85	1.82	1.79
23	2.09	2.08	2.06	2.05	2.04	2.02	2.01	2.01	2.00	1.96	1.91	1.88	1.86	1.84	1.82	1.79	1.76
24	2.07	2.05	2.04	2.03	2.01	2.00	1.99	1.98	1.97	1.94	1.89	1.86	1.84	1.82	1.80	1.77	1.74
25	2.05	2.04	2.02	2.01	2.00	1.98	1.97	1.96	1.96	1.92	1.87	1.84	1.82	1.80	1.78	1.75	1.72
30	1.98	1.96	1.95	1.93	1.92	1.91	1.90	1.89	1.88	1.84	1.79	1.76	1.74	1.72	1.70	1.66	1.63
40	1.89	1.87	1.85	1.84	1.83	1.81	1.80	1.79	1.78	1.74	1.69	1.66	1.64	1.61	1.59	1.55	1.52
50	1.83	1.81	1.80	1.78	1.77	1.76	1.75	1.74	1.73	1.69	1.63	1.60	1.58	1.55	1.52	1.48	1.45
60	1.80	1.78	1.76	1.75	1.73	1.72	1.71	1.70	1.69	1.65	1.59	1.56	1.53	1.51	1.48	1.44	1.40
75	1.76	1.74	1.73	1.71	1.70	1.69	1.67	1.66	1.65	1.61	1.55	1.52	1.49	1.47	1.44	1.39	1.35
100	1.73	1.71	1.69	1.68	1.66	1.65	1.64	1.63	1.62	1.57	1.52	1.48	1.45	1.42	1.39	1.34	1.30
200	1.67	1.66	1.64	1.62	1.61	1.60	1.58	1.57	1.56	1.52	1.46	1.41	1.39	1.35	1.32	1.26	1.21
1000	1.63	1.61	1.60	1.58	1.57	1.55	1.54	1.53	1.52	1.47	1.41	1.36	1.33	1.30	1.26	1.19	1.11

Appendix 8
F Ratio Critical Values

α = 0.025

d.f. denominator	d.f. numerator 1	2	3	4	5	6	7	8	9	10	11	12	13	14	15	16
1	647.79	799.48	864.15	899.60	921.83	937.11	948.20	956.64	963.28	968.63	973.03	976.72	979.84	982.55	984.87	986.91
2	38.51	39.00	39.17	39.25	39.30	39.33	39.36	39.37	39.39	39.40	39.41	39.41	39.42	39.43	39.43	39.44
3	17.44	16.04	15.44	15.10	14.88	14.73	14.62	14.54	14.47	14.42	14.37	14.34	14.30	14.28	14.25	14.23
4	12.22	10.65	9.98	9.60	9.36	9.20	9.07	8.98	8.90	8.84	8.79	8.75	8.72	8.68	8.66	8.63
5	10.01	8.43	7.76	7.39	7.15	6.98	6.85	6.76	6.68	6.62	6.57	6.52	6.49	6.46	6.43	6.40
6	8.81	7.26	6.60	6.23	5.99	5.82	5.70	5.60	5.52	5.46	5.41	5.37	5.33	5.30	5.27	5.24
7	8.07	6.54	5.89	5.52	5.29	5.12	4.99	4.90	4.82	4.76	4.71	4.67	4.63	4.60	4.57	4.54
8	7.57	6.06	5.42	5.05	4.82	4.65	4.53	4.43	4.36	4.30	4.24	4.20	4.16	4.13	4.10	4.08
9	7.21	5.71	5.08	4.72	4.48	4.32	4.20	4.10	4.03	3.96	3.91	3.87	3.83	3.80	3.77	3.74
10	6.94	5.46	4.83	4.47	4.24	4.07	3.95	3.85	3.78	3.72	3.66	3.62	3.58	3.55	3.52	3.50
11	6.72	5.26	4.63	4.28	4.04	3.88	3.76	3.66	3.59	3.53	3.47	3.43	3.39	3.36	3.33	3.30
12	6.55	5.10	4.47	4.12	3.89	3.73	3.61	3.51	3.44	3.37	3.32	3.28	3.24	3.21	3.18	3.15
13	6.41	4.97	4.35	4.00	3.77	3.60	3.48	3.39	3.31	3.25	3.20	3.15	3.12	3.08	3.05	3.03
14	6.30	4.86	4.24	3.89	3.66	3.50	3.38	3.29	3.21	3.15	3.09	3.05	3.01	2.98	2.95	2.92
15	6.20	4.77	4.15	3.80	3.58	3.41	3.29	3.20	3.12	3.06	3.01	2.96	2.92	2.89	2.86	2.84
16	6.12	4.69	4.08	3.73	3.50	3.34	3.22	3.12	3.05	2.99	2.93	2.89	2.85	2.82	2.79	2.76
17	6.04	4.62	4.01	3.66	3.44	3.28	3.16	3.06	2.98	2.92	2.87	2.82	2.79	2.75	2.72	2.70
18	5.98	4.56	3.95	3.61	3.38	3.22	3.10	3.01	2.93	2.87	2.81	2.77	2.73	2.70	2.67	2.64
19	5.92	4.51	3.90	3.56	3.33	3.17	3.05	2.96	2.88	2.82	2.76	2.72	2.68	2.65	2.62	2.59
20	5.87	4.46	3.86	3.51	3.29	3.13	3.01	2.91	2.84	2.77	2.72	2.68	2.64	2.60	2.57	2.55
22	5.79	4.38	3.78	3.44	3.22	3.05	2.93	2.84	2.76	2.70	2.65	2.60	2.56	2.53	2.50	2.47
23	5.75	4.35	3.75	3.41	3.18	3.02	2.90	2.81	2.73	2.67	2.62	2.57	2.53	2.50	2.47	2.44
24	5.72	4.32	3.72	3.38	3.15	2.99	2.87	2.78	2.70	2.64	2.59	2.54	2.50	2.47	2.44	2.41
25	5.69	4.29	3.69	3.35	3.13	2.97	2.85	2.75	2.68	2.61	2.56	2.51	2.48	2.44	2.41	2.38
30	5.57	4.18	3.59	3.25	3.03	2.87	2.75	2.65	2.57	2.51	2.46	2.41	2.37	2.34	2.31	2.28
40	5.42	4.05	3.46	3.13	2.90	2.74	2.62	2.53	2.45	2.39	2.33	2.29	2.25	2.21	2.18	2.15
50	5.34	3.97	3.39	3.05	2.83	2.67	2.55	2.46	2.38	2.32	2.26	2.22	2.18	2.14	2.11	2.08
60	5.29	3.93	3.34	3.01	2.79	2.63	2.51	2.41	2.33	2.27	2.22	2.17	2.13	2.09	2.06	2.03
75	5.23	3.88	3.30	2.96	2.74	2.58	2.46	2.37	2.29	2.22	2.17	2.12	2.08	2.05	2.01	1.99
100	5.18	3.83	3.25	2.92	2.70	2.54	2.42	2.32	2.24	2.18	2.12	2.08	2.04	2.00	1.97	1.94
200	5.10	3.76	3.18	2.85	2.63	2.47	2.35	2.26	2.18	2.11	2.06	2.01	1.97	1.93	1.90	1.87
1000	5.04	3.70	3.13	2.80	2.58	2.42	2.30	2.20	2.13	2.06	2.01	1.96	1.92	1.88	1.85	1.82

Appendix 8
F Ratio Critical Values

α = 0.025

d.f. denominator	d.f. numerator 17	18	19	20	21	22	23	24	25	30	40	50	60	75	100	200	1000
1	988.72	990.35	991.80	993.08	994.30	995.35	996.34	997.27	998.09	1001.4	1005.6	1008.1	1009.8	1011.5	1013.2	1015.7	1017.8
2	39.44	39.44	39.45	39.45	39.45	39.45	39.45	39.46	39.46	39.46	39.47	39.48	39.48	39.48	39.49	39.49	39.50
3	14.21	14.20	14.18	14.17	14.16	14.14	14.13	14.12	14.12	14.08	14.04	14.01	13.99	13.97	13.96	13.93	13.91
4	8.61	8.59	8.58	8.56	8.55	8.53	8.52	8.51	8.50	8.46	8.41	8.38	8.36	8.34	8.32	8.29	8.26
5	6.38	6.36	6.34	6.33	6.31	6.30	6.29	6.28	6.27	6.23	6.18	6.14	6.12	6.10	6.08	6.05	6.02
6	5.22	5.20	5.18	5.17	5.15	5.14	5.13	5.12	5.11	5.07	5.01	4.98	4.96	4.94	4.92	4.88	4.86
7	4.52	4.50	4.48	4.47	4.45	4.44	4.43	4.41	4.40	4.36	4.31	4.28	4.25	4.23	4.21	4.18	4.15
8	4.05	4.03	4.02	4.00	3.98	3.97	3.96	3.95	3.94	3.89	3.84	3.81	3.78	3.76	3.74	3.70	3.68
9	3.72	3.70	3.68	3.67	3.65	3.64	3.63	3.61	3.60	3.56	3.51	3.47	3.45	3.43	3.40	3.37	3.34
10	3.47	3.45	3.44	3.42	3.40	3.39	3.38	3.37	3.35	3.31	3.26	3.22	3.20	3.18	3.15	3.12	3.09
11	3.28	3.26	3.24	3.23	3.21	3.20	3.18	3.17	3.16	3.12	3.06	3.03	3.00	2.98	2.96	2.92	2.89
12	3.13	3.11	3.09	3.07	3.06	3.04	3.03	3.02	3.01	2.96	2.91	2.87	2.85	2.82	2.80	2.76	2.73
13	3.00	2.98	2.96	2.95	2.93	2.92	2.91	2.89	2.88	2.84	2.78	2.74	2.72	2.70	2.67	2.63	2.60
14	2.90	2.88	2.86	2.84	2.83	2.81	2.80	2.79	2.78	2.73	2.67	2.64	2.61	2.59	2.56	2.53	2.50
15	2.81	2.79	2.77	2.76	2.74	2.73	2.71	2.70	2.69	2.64	2.59	2.55	2.52	2.50	2.47	2.44	2.40
16	2.74	2.72	2.70	2.68	2.67	2.65	2.64	2.63	2.61	2.57	2.51	2.47	2.45	2.42	2.40	2.36	2.32
17	2.67	2.65	2.63	2.62	2.60	2.59	2.57	2.56	2.55	2.50	2.44	2.41	2.38	2.35	2.33	2.29	2.26
18	2.62	2.60	2.58	2.56	2.54	2.53	2.52	2.50	2.49	2.44	2.38	2.35	2.32	2.30	2.27	2.23	2.20
19	2.57	2.55	2.53	2.51	2.49	2.48	2.46	2.45	2.44	2.39	2.33	2.30	2.27	2.24	2.22	2.18	2.14
20	2.52	2.50	2.48	2.46	2.45	2.43	2.42	2.41	2.40	2.35	2.29	2.25	2.22	2.20	2.17	2.13	2.09
22	2.45	2.43	2.41	2.39	2.37	2.36	2.34	2.33	2.32	2.27	2.21	2.17	2.14	2.12	2.09	2.05	2.01
23	2.42	2.39	2.37	2.36	2.34	2.33	2.31	2.30	2.29	2.24	2.18	2.14	2.11	2.08	2.06	2.01	1.98
24	2.39	2.36	2.35	2.33	2.31	2.30	2.28	2.27	2.26	2.21	2.15	2.11	2.08	2.05	2.02	1.98	1.94
25	2.36	2.34	2.32	2.30	2.28	2.27	2.26	2.24	2.23	2.18	2.12	2.08	2.05	2.02	2.00	1.95	1.91
30	2.26	2.23	2.21	2.20	2.18	2.16	2.15	2.14	2.12	2.07	2.01	1.97	1.94	1.91	1.88	1.84	1.80
40	2.13	2.11	2.09	2.07	2.05	2.03	2.02	2.01	1.99	1.94	1.88	1.83	1.80	1.77	1.74	1.69	1.65
50	2.06	2.03	2.01	1.99	1.98	1.96	1.95	1.93	1.92	1.87	1.80	1.75	1.72	1.69	1.66	1.60	1.56
60	2.01	1.98	1.96	1.94	1.93	1.91	1.90	1.88	1.87	1.82	1.74	1.70	1.67	1.63	1.60	1.54	1.49
75	1.96	1.94	1.92	1.90	1.88	1.86	1.85	1.83	1.82	1.76	1.69	1.65	1.61	1.58	1.54	1.48	1.43
100	1.91	1.89	1.87	1.85	1.83	1.81	1.80	1.78	1.77	1.71	1.64	1.59	1.56	1.52	1.48	1.42	1.36
200	1.84	1.82	1.80	1.78	1.76	1.74	1.73	1.71	1.70	1.64	1.56	1.51	1.47	1.44	1.39	1.32	1.25
1000	1.79	1.77	1.74	1.72	1.70	1.69	1.67	1.65	1.64	1.58	1.50	1.45	1.41	1.36	1.32	1.23	1.13

Appendix 8
F Ratio Critical Values

α = 0.010

d.f. denominator	d.f. numerator 1	2	3	4	5	6	7	8	9	10	11	12	13	14	15	16
1	4052	4999	5404	5624	5764	5859	5928	5981	6022	6056	6083	6107	6126	6143	6157	6170
2	98.50	99.00	99.16	99.25	99.30	99.33	99.36	99.38	99.39	99.40	99.41	99.42	99.42	99.43	99.43	99.44
3	34.12	30.82	29.46	28.71	28.24	27.91	27.67	27.49	27.34	27.23	27.13	27.05	26.98	26.92	26.87	26.83
4	21.20	18.00	16.69	15.98	15.52	15.21	14.98	14.80	14.66	14.55	14.45	14.37	14.31	14.25	14.20	14.15
5	16.26	13.27	12.06	11.39	10.97	10.67	10.46	10.29	10.16	10.05	9.96	9.89	9.82	9.77	9.72	9.68
6	13.75	10.92	9.78	9.15	8.75	8.47	8.26	8.10	7.98	7.87	7.79	7.72	7.66	7.60	7.56	7.52
7	12.25	9.55	8.45	7.85	7.46	7.19	6.99	6.84	6.72	6.62	6.54	6.47	6.41	6.36	6.31	6.28
8	11.26	8.65	7.59	7.01	6.63	6.37	6.18	6.03	5.91	5.81	5.73	5.67	5.61	5.56	5.52	5.48
9	10.56	8.02	6.99	6.42	6.06	5.80	5.61	5.47	5.35	5.26	5.18	5.11	5.05	5.01	4.96	4.92
10	10.04	7.56	6.55	5.99	5.64	5.39	5.20	5.06	4.94	4.85	4.77	4.71	4.65	4.60	4.56	4.52
11	9.65	7.21	6.22	5.67	5.32	5.07	4.89	4.74	4.63	4.54	4.46	4.40	4.34	4.29	4.25	4.21
12	9.33	6.93	5.95	5.41	5.06	4.82	4.64	4.50	4.39	4.30	4.22	4.16	4.10	4.05	4.01	3.97
13	9.07	6.70	5.74	5.21	4.86	4.62	4.44	4.30	4.19	4.10	4.02	3.96	3.91	3.86	3.82	3.78
14	8.86	6.51	5.56	5.04	4.69	4.46	4.28	4.14	4.03	3.94	3.86	3.80	3.75	3.70	3.66	3.62
15	8.68	6.36	5.42	4.89	4.56	4.32	4.14	4.00	3.89	3.80	3.73	3.67	3.61	3.56	3.52	3.49
16	8.53	6.23	5.29	4.77	4.44	4.20	4.03	3.89	3.78	3.69	3.62	3.55	3.50	3.45	3.41	3.37
17	8.40	6.11	5.19	4.67	4.34	4.10	3.93	3.79	3.68	3.59	3.52	3.46	3.40	3.35	3.31	3.27
18	8.29	6.01	5.09	4.58	4.25	4.01	3.84	3.71	3.60	3.51	3.43	3.37	3.32	3.27	3.23	3.19
19	8.18	5.93	5.01	4.50	4.17	3.94	3.77	3.63	3.52	3.43	3.36	3.30	3.24	3.19	3.15	3.12
20	8.10	5.85	4.94	4.43	4.10	3.87	3.70	3.56	3.46	3.37	3.29	3.23	3.18	3.13	3.09	3.05
22	7.95	5.72	4.82	4.31	3.99	3.76	3.59	3.45	3.35	3.26	3.18	3.12	3.07	3.02	2.98	2.94
23	7.88	5.66	4.76	4.26	3.94	3.71	3.54	3.41	3.30	3.21	3.14	3.07	3.02	2.97	2.93	2.89
24	7.82	5.61	4.72	4.22	3.90	3.67	3.50	3.36	3.26	3.17	3.09	3.03	2.98	2.93	2.89	2.85
25	7.77	5.57	4.68	4.18	3.85	3.63	3.46	3.32	3.22	3.13	3.06	2.99	2.94	2.89	2.85	2.81
30	7.56	5.39	4.51	4.02	3.70	3.47	3.30	3.17	3.07	2.98	2.91	2.84	2.79	2.74	2.70	2.66
40	7.31	5.18	4.31	3.83	3.51	3.29	3.12	2.99	2.89	2.80	2.73	2.66	2.61	2.56	2.52	2.48
50	7.17	5.06	4.20	3.72	3.41	3.19	3.02	2.89	2.78	2.70	2.63	2.56	2.51	2.46	2.42	2.38
60	7.08	4.98	4.13	3.65	3.34	3.12	2.95	2.82	2.72	2.63	2.56	2.50	2.44	2.39	2.35	2.31
75	6.99	4.90	4.05	3.58	3.27	3.05	2.89	2.76	2.65	2.57	2.49	2.43	2.38	2.33	2.29	2.25
100	6.90	4.82	3.98	3.51	3.21	2.99	2.82	2.69	2.59	2.50	2.43	2.37	2.31	2.27	2.22	2.19
200	6.76	4.71	3.88	3.41	3.11	2.89	2.73	2.60	2.50	2.41	2.34	2.27	2.22	2.17	2.13	2.09
1000	6.66	4.63	3.80	3.34	3.04	2.82	2.66	2.53	2.43	2.34	2.27	2.20	2.15	2.10	2.06	2.02

Appendix 8
F Ratio Critical Values

$\alpha = 0.010$

d.f. denominator	d.f. numerator 17	18	19	20	21	22	23	24	25	30	40	50	60	75	100	200	1000
1	6181	6191	6201	6209	6216	6223	6229	6234	6240	6260	6286	6302	6313	6324	6334	6350	6363
2	99.44	99.44	99.45	99.45	99.45	99.46	99.46	99.46	99.46	99.47	99.48	99.48	99.48	99.48	99.49	99.49	99.50
3	26.79	26.75	26.72	26.69	26.66	26.64	26.62	26.60	26.58	26.50	26.41	26.35	26.32	26.28	26.24	26.18	26.14
4	14.11	14.08	14.05	14.02	13.99	13.97	13.95	13.93	13.91	13.84	13.75	13.69	13.65	13.61	13.58	13.52	13.47
5	9.64	9.61	9.58	9.55	9.53	9.51	9.49	9.47	9.45	9.38	9.29	9.24	9.20	9.17	9.13	9.08	9.03
6	7.48	7.45	7.42	7.40	7.37	7.35	7.33	7.31	7.30	7.23	7.14	7.09	7.06	7.02	6.99	6.93	6.89
7	6.24	6.21	6.18	6.16	6.13	6.11	6.09	6.07	6.06	5.99	5.91	5.86	5.82	5.79	5.75	5.70	5.66
8	5.44	5.41	5.38	5.36	5.34	5.32	5.30	5.28	5.26	5.20	5.12	5.07	5.03	5.00	4.96	4.91	4.87
9	4.89	4.86	4.83	4.81	4.79	4.77	4.75	4.73	4.71	4.65	4.57	4.52	4.48	4.45	4.41	4.36	4.32
10	4.49	4.46	4.43	4.41	4.38	4.36	4.34	4.33	4.31	4.25	4.17	4.12	4.08	4.05	4.01	3.96	3.92
11	4.18	4.15	4.12	4.10	4.08	4.06	4.04	4.02	4.01	3.94	3.86	3.81	3.78	3.74	3.71	3.66	3.61
12	3.94	3.91	3.88	3.86	3.84	3.82	3.80	3.78	3.76	3.70	3.62	3.57	3.54	3.50	3.47	3.41	3.37
13	3.75	3.72	3.69	3.66	3.64	3.62	3.60	3.59	3.57	3.51	3.43	3.38	3.34	3.31	3.27	3.22	3.18
14	3.59	3.56	3.53	3.51	3.48	3.46	3.44	3.43	3.41	3.35	3.27	3.22	3.18	3.15	3.11	3.06	3.02
15	3.45	3.42	3.40	3.37	3.35	3.33	3.31	3.29	3.28	3.21	3.13	3.08	3.05	3.01	2.98	2.92	2.88
16	3.34	3.31	3.28	3.26	3.24	3.22	3.20	3.18	3.16	3.10	3.02	2.97	2.93	2.90	2.86	2.81	2.76
17	3.24	3.21	3.19	3.16	3.14	3.12	3.10	3.08	3.07	3.00	2.92	2.87	2.83	2.80	2.76	2.71	2.66
18	3.16	3.13	3.10	3.08	3.05	3.03	3.02	3.00	2.98	2.92	2.84	2.78	2.75	2.71	2.68	2.62	2.58
19	3.08	3.05	3.03	3.00	2.98	2.96	2.94	2.92	2.91	2.84	2.76	2.71	2.67	2.64	2.60	2.55	2.50
20	3.02	2.99	2.96	2.94	2.92	2.90	2.88	2.86	2.84	2.78	2.69	2.64	2.61	2.57	2.54	2.48	2.43
22	2.91	2.88	2.85	2.83	2.81	2.78	2.77	2.75	2.73	2.67	2.58	2.53	2.50	2.46	2.42	2.36	2.32
23	2.86	2.83	2.80	2.78	2.76	2.74	2.72	2.70	2.69	2.62	2.54	2.48	2.45	2.41	2.37	2.32	2.27
24	2.82	2.79	2.76	2.74	2.72	2.70	2.68	2.66	2.64	2.58	2.49	2.44	2.40	2.37	2.33	2.27	2.22
25	2.78	2.75	2.72	2.70	2.68	2.66	2.64	2.62	2.60	2.54	2.45	2.40	2.36	2.33	2.29	2.23	2.18
30	2.63	2.60	2.57	2.55	2.53	2.51	2.49	2.47	2.45	2.39	2.30	2.25	2.21	2.17	2.13	2.07	2.02
40	2.45	2.42	2.39	2.37	2.35	2.33	2.31	2.29	2.27	2.20	2.11	2.06	2.02	1.98	1.94	1.87	1.82
50	2.35	2.32	2.29	2.27	2.24	2.22	2.20	2.18	2.17	2.10	2.01	1.95	1.91	1.87	1.82	1.76	1.70
60	2.28	2.25	2.22	2.20	2.17	2.15	2.13	2.12	2.10	2.03	1.94	1.88	1.84	1.79	1.75	1.68	1.62
75	2.22	2.18	2.16	2.13	2.11	2.09	2.07	2.05	2.03	1.96	1.87	1.81	1.76	1.72	1.67	1.60	1.53
100	2.15	2.12	2.09	2.07	2.04	2.02	2.00	1.98	1.97	1.89	1.80	1.74	1.69	1.65	1.60	1.52	1.45
200	2.06	2.03	2.00	1.97	1.95	1.93	1.90	1.89	1.87	1.79	1.69	1.63	1.58	1.53	1.48	1.39	1.30
1000	1.98	1.95	1.92	1.90	1.87	1.85	1.83	1.81	1.79	1.72	1.61	1.54	1.50	1.44	1.38	1.28	1.16

Index

Notes

Notes

Notes

Notes

About the Author

Gerald C. (Jerry) Swanson was taught statistics as an undergraduate Chemistry major in the late 1960's. (While getting a good grade, he did not really learn useful statistics.) He later worked at the Los Alamos National Laboratory as a nuclear analytical chemist. There he learned statistics because a statistician was able to translate it into the language of chemistry. (He also completed an M.S. and Ph.D. in Nuclear Analytical Chemistry.)

In the late 1970"s he attended a cutting-edge M.A. program in the Applied Behavioral Sciences preparatory to a career change as an Organization Development consultant. As part of the program, he began to translate his knowledge of statistics into the language of the behavioral sciences. He further developed his experiential teaching methods while teaching statistics in an M.A. Psychology program at Antioch University in Seattle in the early 1980's.

The contents of this book and its teaching approach were further developed in the middle 1990's when Jerry began teaching statistics as an introductory course in a B.S. Applied Behavioral Science program and at the graduate level in the M.A. Applied Behavioral Sciences programs thorough the Leadership Institute of Seattle at Bastyr University in the Seattle area. The content, sequence and methods of introducing the materials have been honed through teaching hundreds of students who would describe themselves as "mathematically challenged". Jerry continues to teach at the graduate level and also delivers three-day intensives covering the entirety of introductory statistics. He may be reached via e-mail at information@greyheronpress.com. Information about the programs mentioned may be found at www.lios.org and at www.bastyr.edu.

Jerry is an internal Organization Development consultant in The Boeing Company in the Seattle area.